STUDIES IN ALGEBRAIC LOGIC

Studies in Mathematics

The Mathematical Association of America

Volume 1: STUDIES IN MODERN ANALYSIS
edited by R. C. Buck

Volume 2: STUDIES IN MODERN ALGEBRA
edited by A. A. Albert

Volume 3: STUDIES IN REAL AND COMPLEX ANALYSIS
edited by I. I. Hirschman, Jr.

Volume 4: STUDIES IN GLOBAL GEOMETRY AND ANALYSIS
edited by S. S. Chern

Volume 5: STUDIES IN MODERN TOPOLOGY
edited by P. J. Hilton

Volume 6: STUDIES IN NUMBER THEORY
edited by W. J. Le Veque

Volume 7: STUDIES IN APPLIED MATHEMATICS
edited by A. H. Taub

Volume 8: STUDIES IN MODEL THEORY
edited by M. D. Morley

Volume 9: STUDIES IN ALGEBRAIC LOGIC
edited by Aubert Daigneault

William Craig
University of California, Berkeley

Aubert Daigneault
Université de Montréal

J. Donald Monk
University of Colorado, Boulder

Helena Rasiowa
University of Warsaw

Gonzalo E. Reyes
Université de Montréal

Studies in Mathematics

Volume 9

STUDIES IN ALGEBRAIC LOGIC

Aubert Daigneault, editor
Université de Montréal

Published and distributed by
The Mathematical Association of America

Library of Congress Catalog Card Number 74-84580

Complete Set ISBN 0-88385-100-8
Vol. 9 ISBN 0-88385-109-1

Printed in the United States of America

Current printing (last digit)

10 9 8 7 6 5 4 3 2 1

CONTENTS

INTRODUCTION

Algebraic logic is the study of algebraic structures arising in mathematical logic or of generalizations of such structures and, insofar as such structures are pertinent, the study of logical problems by algebraic means. It all started with Boole's "An Investigation of the Laws of Thought" first published in 1854. This was the beginning of the theory of Boolean algebras which is the algebraic counterpart to classical propositional logic. This early "algebra of logic," as it was then called, received a comprehensive treatment in Schröder's "Vorlesungen über die Algebra der Logik" published from 1890 till 1905. Mostly with the works of Stone and Tarski in the nineteen-thirties the theory of Boolean algebras became well established as a chapter of abstract algebra.

Through various intermediate stages, the process of the algebraization of logic led to structures called polyadic and cylindric algebras in the nineteen-fifties. Roughly these are Boolean algebras with some operators corresponding to quantifiers or substitutions. The first example of such a structure is that of all equivalence classes of formulas in a first order theory. The phrase "algebraic logic" appears to have been coined by Halmos in his expository paper "The Basic Concepts of Algebraic Logic" (American Mathematical Monthly, volume 63, 1956). This article by Halmos and others by him have been brought together in a book entitled *Algebraic Logic* (Chelsea, 1962) and which remains the main basic reference work on the theory of polyadic algebras. The first part of a monumental two-part *Cylindric Algebras* by Henkin, Monk and Tarski has recently appeared (North Holland, 1971).

As will be seen from the four papers brought under this cover, algebraic logic did not stop with the invention of polyadic and cylindric algebras and the development of their theory. These papers which were all written especially for this Study present rather different facets of contemporary concepts and results in the subject.

Craig's article is concerned with an axiomatization of the theory of polyadic algebras with equality (or with generalized diagonal elements when non-finite transformations are essentially involved) that is based solely on transformations, Boolean operations, and diagonal elements as opposed to one that uses also quantifiers. For each transformation α there is in the ordinary definition of polyadic algebra the substitution operator $s\alpha$ which is an inverse image operator. The author introduces also a direct image operator $t\alpha$. A Boolean algebra which, for every α taken from a semigroup of transformations on the set of variables or, indeed, from an abstract semigroup, is equipped with operators $s\alpha$ and $t\alpha$ in such a way that certain equations are satisfied, is called a trans-Boolean algebra. The main result, Theorem 29, asserts that under certain conditions on the semigroup of transformations, the concept of polyadic algebra with diagonal elements and adapted to this semigroup is equivalent to that of trans-Boolean algebra belonging to that semigroup. This shows that it is possible to give priority to transformation operators over quantifiers contrary to what is done in the theory of cylindric algebras.

Logical structures sometimes arise in contexts quite remote from that of formal logic. Monk's paper gives three situations in combinatorial theory where (mostly finite dimensional) cylindric algebras show up unexpectedly and determines the significance of the representability of these algebras in those situations. In the first instance, he associates to a quasi-group, in particular to a loop, a 3-dimensional cylindric algebra. He shows that the loop is a group if and only if the algebra is representable.

The second example is drawn from a geometrical context. Indeed, an intuitively motivated association of a 3-dimensional cylindric algebra A to a projective geometry G is made. It turns out that if G is a hyperplane in a space of one higher dimension, then A is representable, and that the converse also holds in case G is finite. In particular, if G is finite and has dimension 1, then G is a line in a projective plane if and only if A is representable. Using a known theorem in geometry, this yields infinitely many examples of non-representable 3-dimensional cylindric algebras. In the third example, a link is established between the representability of some cylindric algebras and the computation of certain numbers in a purely combinatorial theorem of Ramsey.

The contribution of Rasiowa to this study describes the connections between Post algebras on the one hand and propositional and first order m-valued logics on the other. First, she establishes in a very exact manner the relationships between various propositional calculi and the corresponding algebras: the classical calculus and Boolean algebras; the intuitionistic calculus and pseudo-Boolean algebras, also known as Heyting algebras; the linear intuitionistic calculus and linear pseudo-Boolean algebras; the minimal propositional calculus and μ-algebras; the positive propositional calculus and relatively pseudo-complemented lattices; the modal propositional calculus of Lewis and topological Boolean algebras, etc. After treating also the case of the m-valued propositional calculus and Post algebras of order m, the author passes to first order m-valued logic. Several concepts and results of classical first order logic are generalized to this many valued logic: the notion of ultraproduct of interpretations and its fundamental property, the compactness theorem, Herbrand's theorem, Hilbert's second ϵ-theorem, Craig's theorem, etc.

In his Ph.D. thesis written in 1967 at the University of Alberta, Edmonton, Alberta, Canada, and entitled "Polyadic Post Algebras," V. J. Willis generalized the concept of polyadic algebra to obtain an algebraic structure that bears to the m-valued first order calculus the same relationship born by ordinary polyadic algebras to the classical first order calculus.

In recent years a new approach to algebraic logic has been initiated mostly by F. W. Lawvere which might be called "categorical logic." The latest craze in that area centers around the concept of topos. This is the central concept in Reyes' article.

The category of all sheaves on a topological space is the simplest example of a topos. An important special case is the category of sets which is the category of sheaves over a one-point space. Replacing the (category of) open sets of the topological space by an arbitrary small category, and generalizing, under the name (Grothendieck) topology, the operator "cover" which, to an open set, associates the class of its coverings by open sets, Grothendieck has shown that the construction of a sheaf still carries over. This gives rise to the concept of a (Grothendieck) topos. The propositional connectives and the first order quantifiers are operations that make sense in a topos, though their definitions are geometrical rather than logical. All this is explained in Reyes' paper which presupposes little knowledge of category theory. As explained by Freyd in "Aspects of topoi" (Bull. Austral. Math. Soc. Vol. 7 (1972) pp. 1–76), it is even possible to define the first order calculus of a topos.

The theme of the paper is a continuing search for unifying "the geometric and logical aspects of a Grothendieck topos." The author first deals with regular categories. In these, the logical operations of conjunction, substitution, and existential quantification are possible. Then he "enriches" the situation by bringing in disjunction, universal quantification, and implication thus obtaining Heyting categories which reflect intuitionistic rather than classical logic.

There is an obvious forgetful functor from the category of topoi to that of Heyting categories. The author looks in vain for a left adjoint to this functor. Unable to do away altogether with geometry, he settles for one in the case of the forgetful functor from

Boolean topoi to Boolean categories with continuous Boolean functors. According to some work of D. Higgs, this gives, in a particular case, a new construction of Boolean powers.

In the last chapter, three categorical versions of logical results are presented. The first two, due to A. Joyal, pertain, one to the characterization of limit ultrapowers, and the other to a "completeness theorem" for Boolean categories. The last one tells us what the notion of enlargement becomes in this context.

AUBERT DAIGNEAULT

UNIFICATION AND ABSTRACTION IN ALGEBRAIC LOGIC*

William Craig

SUMMARY

Algebraic logic is concerned with operations on sets of sequences. If the sequences are of length 1, then one often considers only (*set-theoretic*) *Boolean operations*, i.e., operations generated by substitution from complementation, union, and intersection. If there are sequences of other length, then one usually considers

* Most of the results of this paper were obtained during 1968–69, when I was on a sabbatical leave from the University of California and also held a post-doctoral fellowship from the National Science Foundation. Other results, including the abstraction from semigroups of transformations to arbitrary semigroups, have been obtained during a subsequent period of support by grant GP-26261 of the National Science Foundation. I should like to thank Mr. Peter Eggenberger and Mr. Douglas J. Young, Sr. for assistance in checking proofs and for helpful suggestions. I should also like to thank Professor Charles Pinter for valuable comments on an earlier draft.

additional operations. Quite often, as we shall see, these arise from one additional source.

To be a little more specific, let L be a language for an algebraic theory of logic. The set-theoretic operations expressible (by means of a term) in L are generated by substitution from certain operations chosen as primitive. In the majority of cases from the literature, one takes a single set I, limits oneself to sequences of length I (i.e., to sequences that are indexed by I), and then chooses as primitive certain operations on sets of such sequences. Now let each mapping of I into I be a *transformation on* I, and let $\mathcal{S}$ be any set of such transformations. As we shall see, $\mathcal{S}$ gives rise in a natural way to certain operations on sets of sequences of length I. Call these the (*set-theoretic*) *transformational operations* (*resulting from* $\mathcal{S}$). For several L from the literature we shall give an $\mathcal{S}$ such that the set-theoretic transformational operations resulting from $\mathcal{S}$ and the set-theoretic Boolean operations together generate by substitution exactly the set-theoretic operations expressible in L. Except for the choice of primitives, these L are therefore characterized by a "parameter" $\mathcal{S}$. This is one unification we have in mind.

For each semigroup $\mathcal{S}$, an algebraic theory $TB_{\mathcal{S}}$ concerning these operations will also be given. The primitives are certain abstract Boolean operations and the abstract counterparts of those set-theoretic transformational operations which are most directly related to $\mathcal{S}$. These counterparts fall into two classes, each operation of one class being definable by a simple first-order formula in terms of Boolean operations and an operation of the other class. Thus, there is a related second, internal, unification.

For each semigroup $\mathcal{S}$ of transformations on I, those changes in the theory of Halmos [4], pp. 213–239, which are indicated in Lucas [9] yield a theory $PD_{\mathcal{S}}$ of polyadic algebras with (generalized) diagonal elements. In contrast to the axiom schemes of $PD_{\mathcal{S}}$, those of $TB_{\mathcal{S}}$ involve only the algebraic structure of $\mathcal{S}$ and are applicable to arbitrary semigroups $\mathcal{S}$. Thus, we also effect a certain abstraction.

We shall also consider the theory $TB_{\mathcal{S}}^{-}$ which results from $TB_{\mathcal{S}}$

when one drops the last axiom scheme. For many semigroups $\mathfrak{S}$ of transformations on I we shall show that $PD_{\mathfrak{S}}$ and $TB_{\mathfrak{S}}^{-}$ are equivalent. There are fewer axiom schemes for $TB_{\mathfrak{S}}^{-}$ than there are for $PD_{\mathfrak{S}}$, and their content seems to be clearer.

Some smaller fragments of $TB_{\mathfrak{S}}$ will also be investigated. One of our aims is to bring out different levels of generality in algebraic logic.

Representation problems of a certain kind will also be considered. By the basic results of Jónsson and Tarski in [7], they will be reduced to related problems for semigroups. Whether this reduction is very helpful remains to be seen.

1. EXTRA-BOOLEAN AND TRANSFORMATIONAL OPERATIONS ON SETS OF SEQUENCES

In §1 we shall consider two classes of operations on sets of sequences and show that they often coincide.

We begin with some conventions, most of which will be used throughout the paper. A *function* shall be a set ϕ of ordered pairs such that if $\langle v, w\rangle$ and $\langle v', w\rangle$ are in ϕ then $v = v'$. The *restriction* of a function ϕ to a set W shall be the set $\{\langle v, w\rangle \in \phi : w \in W\}$, which is also a function. If ϕ and ψ have the same restriction to W, they shall *agree on* W. We let dom ϕ, *the domain of* ϕ, be the set $\{w : \langle v, w\rangle$ is in ϕ for some $v\}$. We let ran ϕ, *the range of* ϕ, be the set $\{v : \langle v, w\rangle$ is in ϕ for some $w\}$. For any sets W and V we let ${}^{W}V = \{\phi : \phi$ is a function, dom $\phi = W$, ran $\phi \subseteq V\}$. Any ϕ in ${}^{W\times W}W = {}^{2W}W$, shall be a 2-*ary operation on* W. Any ϕ in ${}^{W}W$ shall be a 1-*ary operation*, or *transformation, on* W. Clearly, if ϕ and ψ are transformations on W, so is their product under composition. For any $\mathfrak{J} \subseteq {}^{W}W$ we let cl$\mathfrak{J}$, *the closure of* $\mathfrak{J}$ *under composition*, be the set consisting of the elements of $\mathfrak{J}$ and of the finite products of elements of $\mathfrak{J}$. If $\mathfrak{J} = $ cl$\mathfrak{J}$, then $\mathfrak{J}$, together with the composition operation, shall be a *semigroup of transformations on* W. We shall often identify this semigroup with $\mathfrak{J}$.

We take an arbitrary set I and an arbitrary set U and shall keep them fixed, unless indicated otherwise. They will serve as our index set and as our universe of objects respectively. Functions in ${}^{I}U$ shall be *sequences* (*of elements of* U).

Variables will be used as follows: $i, j, k, \cdots$ for elements of I; $J, K, J', \cdots$ for subsets of I; $M, N, M', \cdots$ for binary relations on I, i.e., for subsets of ${}^{2}I = I \times I$; $f, g, h, \cdots$ for sequences. Also, in §1, we let $\alpha, \beta, \gamma, \cdots$ be transformations on I, and $x, y, z, \cdots$ sets of sequences.

We shall use $f(i)$, fi, and f_i to indicate the value of f for the argument i, or the ith *term* of f. Sometimes, for suggestiveness, if $I = \{0, 1, \cdots, n\}$ we let f be $\langle f_0, f_1, \cdots, f_n\rangle$ and if I is the set $\{0, 1, \cdots\}$ of all natural numbers we let f be $\langle f_0, f_1, \cdots\rangle$.

We let $M^{\smile}$ be the converse of M. Formation of the relative product, and hence in particular composition of functions, will be indicated either by $\circ$ or by juxtaposition. Thus, if $I = \{0, 1, \cdots, n\}$, then $f \circ \alpha = f\alpha = \langle f_{\alpha(0)}, f_{\alpha(1)}, \cdots, f_{\alpha(n)}\rangle$. Also $M \circ N = MN = \{\langle i, k\rangle$: there is some j such that $\langle i, j\rangle \in M$ and $\langle j, k\rangle \in N\}$. Hence $\alpha^{\smile}\alpha$ is the equivalence relation $\{\langle i, k\rangle : \alpha(i) = \alpha(k)\}$. Whether a juxtaposition is used to indicate a relative product or the value of a function for an argument should be clear from the context.

Clearly, each $f\alpha$ is a sequence. In §1, we let $\hat{\alpha}$ be the function whose domain is ${}^{I}U$ which maps each f into the sequence $\hat{\alpha}(f) = f\alpha$. Thus, in §1, we regard ^ as a function whose domain is ${}^{I}I$ which maps each α into $\hat{\alpha}$.

We call α *idempotent* if and only if $\alpha\alpha = \alpha$. Thus, α is idempotent if and only if $\alpha(j) = j$ for each $j \in \operatorname{ran} \alpha$. In §1, we let id be the identity transformation on I. It satisfies $\operatorname{id}(j) = j$ for each j. If $\alpha\beta = \operatorname{id}$, then α shall be a *left inverse* of β and β a *right inverse* of α. For any set $\mathcal{S}$ of transformations on I, we let:

$$\mathcal{I}(\mathcal{S}) = \{\alpha \in \operatorname{cl}\mathcal{S} : \alpha\alpha = \alpha\}.$$

$$\mathcal{R}(\mathcal{S}) = \{\alpha \in \operatorname{cl}\mathcal{S} : \beta\alpha = \operatorname{id} \text{ for some } \beta \text{ in } \operatorname{cl}\mathcal{S}\}.$$

$$\mathcal{L}(\mathcal{S}) = \{\alpha \in \operatorname{cl}\mathcal{S} : \alpha\beta = \operatorname{id} \text{ for some } \beta \text{ in } \operatorname{cl}\mathcal{S}\}.$$

We let eqv M be the intersection of the family $\{N : \operatorname{id} + M \subseteq N$ and N is an equivalence relation$\}$. Thus eqv M is the equivalence relation on I which is generated by M. We let $M \upharpoonright J = \text{id} + \{\langle i, k\rangle \in M : i \in J \text{ and } k \in J\}$, and call it the *confinement of M to J*. Note that eqv $M = M$ implies that eqv $(M \upharpoonright J) = M \upharpoonright J$.

We let $\| J \|$ be the cardinality of J. We call J a *support* of α if and only if $\alpha(i) = i$ for each $i \notin J$. For each infinite cardinal κ, we let:

$$\mathcal{S}_\kappa = \{\alpha : \text{there is some } J \text{ supporting } \alpha \text{ such that } \| J \| < \kappa\}.$$

$$\mathcal{C}_\kappa = \{J : \| J \| < \kappa\}.$$

$$\mathcal{E}_\kappa = \{M : M = \text{eqv}\, M \text{ and } \| \{i : \langle i, j\rangle \in M \text{ for some } j \neq i\} \| < \kappa\}.$$

For any i and k we let (i/k), *the replacement of i by k*, be the transformation on I which satisfies $(i/k)(i) = k$ and $(i/k)(j) = j$ for $i \neq j$. For any i and k we let (i, k), *the transposition of i and k*, be the transformation on I which satisfies $(i, k)(i) = k$, $(i, k)(k) = i$, and $(i, k)(j) = j$ for $i \neq j \neq k$. Evidently, each replacement and each transposition is in $\mathcal{S}_\omega$, i.e., has some finite support. Also, one sees easily that each γ in $\mathcal{S}_\omega$ which is idempotent is a product of replacements, that each β in $\mathcal{S}_\omega$ which is a right and a left inverse is a product of transpositions, and that each α in $\mathcal{S}_\omega$ is a product $\beta\gamma$ where γ is in $\mathcal{S}_\omega$ and is idempotent and β is in $\mathcal{S}_\omega$ and is a right and a left inverse (cf. [4], p. 146). It follows that $\text{cl}\{\alpha \in \mathcal{S}_\omega : \alpha\alpha = \alpha\} = \text{cl}\{(i/k) : i \in I,\ k \in I\}$ and that $\mathcal{S}_\omega = \text{cl}(\{(i, k) : i \in I, k \in I\} + \{(i/k) : i \in I, k \in I\})$.

We let, for example, $\alpha^*J = \{\alpha(i) : i \in J\}$, $\alpha^*M = \alpha M \alpha^{\smile} = \{\langle\alpha(i), \alpha(k)\rangle : \langle i, k\rangle \in M\}$, $\alpha^{-1}J = \{i : \alpha(i) \in J\}$, and $\alpha^{-1}M = \alpha^{\smile} M \alpha = \{\langle i, k\rangle : \langle\alpha(i), \alpha(k)\rangle \in M\}$.

At last, we turn to operations on sets of sequences. The 2-ary operations of *intersection* and *union* transform any given pair $\langle x, y\rangle$ of sets of sequences into the set $x \cdot y = \{f : f \in x \text{ and } f \in y\}$ and into the set $x + y = \{f : f \in x \text{ or } f \in y \text{ or both}\}$ respectively. The 1-ary operation of *complementation* (with respect to ${}^IU$) transforms any given x into $-x = \{f : f \notin x\}$. The (*set-theoretic*) *Boolean operations* shall be generated from these three operations by substitution and shall include the 0-ary operations $\varnothing = \{f : f \neq f\}$ and $\mathbb{1} = {}^IU$.

Next, for any α, J, and M we define on $\{x : x \subseteq {}^IU\}$ a 1-ary operation $(s\alpha)$, 1-ary operation (cJ), 0-ary operation (dM), and 1-ary operation (eM) as follows:

(1) $(s\alpha)x = \{f : f\alpha \in x\} = \{f : \hat{\alpha}(f) \in x\} = \hat{\alpha}^{-1}x.$

(2) $(cJ)x = \{f$: for some $h \in x, f(i) = h(i)$ for each $i \notin J\}.$

(3) $(dM) = \{f : f(i) = f(k)$ for each $\langle i, k\rangle \in M\}.$

(4) $(eM)x = \{f \in x : f(i) = f(k)$ for each $\langle i, k\rangle \in M\}.$

We call $(s\alpha)$ *pre-image formation with respect to* α, (cJ) *cylindrification with respect to* J, (dM) the *diagonal element determined by* M, and (eM) *diagonalization with respect to* M. When $I = \{0, 1\}$ and U is the set of real numbers, then a sequence $f = \langle f_0, f_1\rangle$ can be thought of as the point in the plane whose horizontal coordinate is f_0 and whose vertical coordinate is f_1. Then $(d\{\langle 0, 1\rangle\})$ is one of the two diagonal lines formed by the points equidistant from the two axes. Also, for each set x of sequences or points, $(c\{0\})x$ is the union of all vertical lines which contain at least one point belonging to x. Then $(c\{0\})$ and $(c\{1\})$ are analogues in 2-dimensional space of operations of cylindrification in 3-dimensional space.

It may be helpful to mention here that there is a correspondence between $(s\alpha)$, (cJ), (dM) and certain symbolic processes of first-order (or elementary) logic. We shall not make use of this correspondence, and refer the reader to [5] for more details. To $(s\alpha)$ there corresponds for $I = \omega$ the process of substituting $v_{\alpha(0)}, v_{\alpha(1)}, \cdots$ for the free occurrences of the individual variables $v_0, v_1, \cdots$ respectively in a first-order formula. When J is $\{i_0, i_1, \cdots, i_k\}$, then to (cJ) there corresponds the process of prefixing $\exists v_{i_0} \exists_{v_{i1}} \cdots \exists_{v_{ik}}$ to a first-order formula. And when M is $\{\langle i_0, i_1\rangle, \langle i_0, i_2\rangle, \langle i_3, i_4\rangle\}$, for example, then to (dM) there corresponds the conjunction $v_{i_0} \doteq v_{i_1} \wedge v_{i_0} \doteq v_{i_2} \wedge v_{i_3} \doteq v_{i_4}$ of equalities.

Largely because of this correspondence, the primitive operations chosen for some algebraic languages of logic are certain Boolean operations, and those operations $(s\alpha)$, (cJ), and (dM) such that α belongs to a certain set $\mathcal{S} \subseteq {}^I I$, J belongs to a certain set $\mathcal{C} \subseteq \{J : J \subseteq I\}$, and M belongs to a certain set $\mathcal{E} \subseteq \{M : M \subseteq I \times I\}$.

For the language of cylindric algebras, which is described in [5] and which aims at economy of primitives, one chooses for $\mathcal{S}$, $\mathcal{C}$, $\mathcal{E}$ respectively the sets $\varnothing$, $\{\{i\} : i \in I\}$, and $\{\{\langle i, k\rangle\} : \langle i, k\rangle \in {}^2I\}$. For the language of polyadic algebras with diagonal elements which is described in Halmos [4], pp. 213–239, one chooses ${}^II$, $\{J : J \subseteq I\}$, and $\{\{\langle i, k\rangle\} : \langle i, k\rangle \in {}^2I\}$. For the modification of this language which occurs in Lucas [9], one chooses for any given infinite cardinal κ the sets $\mathcal{S}_\kappa$, $\mathcal{C}_\kappa$, and $\mathcal{E}_\kappa$.

Given (dM) and $\cdot$, one can define $(eM)x$ as $x \cdot (dM)$. Conversely, given (eM) and $\mathbf{1}$, one can define (dM) as $(eM)\mathbf{1}$. Hence, in general, the set of operations expressible in an algebraic language of logic is not affected if, for each $M \in \mathcal{E}$, one replaces (dM) as primitive by (eM).

Let any $\mathcal{S}$, $\mathcal{C}$, $\mathcal{E}$ be given. By what has just been said, it is natural to consider the sets $\{(s\alpha) : \alpha \in \mathcal{S}\}$, $\{(cJ) : J \in \mathcal{C}\}$, $\{(eM) : M \in \mathcal{E}\}$. Since each of these consists of 1-ary operations, it is also natural to consider the closure under composition of their union $\{(s\alpha) : \alpha \in \mathcal{S}\} + \{(cJ) : J \in \mathcal{C}\} + \{(eM) : M \in \mathcal{E}\}$. The operations in this closure shall be the (*set-theoretic*) *extra-Boolean operations* (*resulting from* $\langle \mathcal{S}, \mathcal{C}, \mathcal{E}\rangle$). We shall also talk about the (set-theoretic) extra-Boolean operations expressible (by means of a term) in a language. For example, given κ, the extra-Boolean operations expressible in Lucas' language are the extra-Boolean operations resulting from $\langle \mathcal{S}_\kappa, \mathcal{C}_\kappa, \mathcal{E}_\kappa\rangle$.

To give an illustration, we now prove the well-known fact that if $i \neq k$ then $(s(i/k)) = (c\{i\})(e\{\langle i, k\rangle\})$. It follows that $(s(i/k))$ is among the extra-Boolean operations expressible in the language of cylindric algebras. For convenience, we shall deal explicitly with that case only where $I = \omega$ and $i < k$. For the other cases the proof is similar. Let any x and any $\langle f_0, \cdots, f_{i-1}, f_i, f_{i+1}, \cdots, f_k, \cdots\rangle = f$ be given. Using (2), (4), and (1) for the first, third, and last step respectively we see that the following conditions are equivalent to each other: f is in $(c\{i\})(e\langle i, k\rangle\})x$; there is some h in $(e\{\langle i, k\rangle\})x$ such that $f(j) = h(j)$ for each $j \notin \{i\}$; there is some $u \in U$ such that $\langle f_0, \cdots, f_{i-1}, u, f_{i+1}, \cdots, f_k, \cdots\rangle$ is in $(e\{\langle i, k\rangle\})x$; there is some $u \in U$ such that $\langle f_0, \cdots, f_{i-1}, u, f_{i+1}, \cdots, f_k, \cdots\rangle$ is in x and $u = f_k$; $\langle f_0, \cdots, f_{i-1}, f_k, f_{i+1}, \cdots, f_k, \cdots\rangle$ is in x; $\langle f_{(i/k)(0)}, f_{(i/k)(1)}, \cdots\rangle$ is in x; $f \circ (i/k)$ is in x; f is in $(s(i/k))x$.

Using (1), (2), and (4) respectively, one verifies easily the following:

(5) $(s\alpha)(s\beta) = (s\alpha\beta)$.

(6) $(cJ)(cK) = (c(J + K))$.

(7) $(eM)(eN) = (e \operatorname{eqv}(M + N))$.

To continue with our illustration, it follows from the last two paragraphs and from $\mathrm{cl}\{\alpha \in \mathcal{S}_\omega : \alpha\alpha = \alpha\} = \mathrm{cl}\{(i/k) : i \in I, k \in I\}$ that the extra-Boolean operations expressible in the language of cylindric algebras result from $\langle \mathrm{cl}\{\alpha \in \mathcal{S}_\omega : \alpha\alpha = \alpha\}, \mathcal{C}_\omega, \mathcal{E}_\omega \rangle$.

Given α, it is natural to consider, along with the pre-image formation $(s\alpha)$, the following operation $(t\alpha)$ of *image formation with respect to* α:

(8) $(t\alpha)x = \{f\alpha : f \in x\} = \{\hat{\alpha}(f) : f \in x\} = \hat{\alpha}^* x$.

Given any set $\mathcal{S}$ of transformations, one is thus led to the union $\{(s\alpha) : \alpha \in \mathcal{S}\} + \{(t\alpha) : \alpha \in \mathcal{S}\}$ and then to the closure of this union under composition. The operations in this closure shall be the (*set-theoretic*) *transformational operations* (*resulting from* $\mathcal{S}$).

Either directly from the definitions involved, or from (13) below and from (5), one sees that the same transformational operations result from $\mathcal{S}$ and from $\mathrm{cl}\mathcal{S}$.

For our purposes the following two identities, which hold for any α, are fundamental. Here, $-\mathrm{ran}\ \alpha = I \cdot -(\mathrm{ran}\ \alpha)$.

(9) $(c - \mathrm{ran}\ \alpha) = (s\alpha)(t\alpha)$.

(10) $(e\alpha^{\smile}\alpha) = (t\alpha)(s\alpha)$.

To verify (9), let any f and x be given. Then, using (1), (8), and (2) for the first, second, and last step respectively, we see that the following conditions are equivalent to each other: f is in $(s\alpha)(t\alpha)x$; $f\alpha$ is in $(t\alpha)x$; there is some h in x such that $h\alpha = f\alpha$; there is some h in x such that $f(i) = h(i)$ for each $i \in \mathrm{ran}\ \alpha$; f is in $(c - \mathrm{ran}\ \alpha)x$.

To verify (10), let any f and x be given. Then, using (8), (1), and (4) for the first, second, and last step respectively, we see that the following conditions are equivalent to each other: f is in $(t\alpha)(s\alpha)x$; there is some h in $(s\alpha)x$ such that $h\alpha = f$; there is some h such that $h\alpha$ is in x and $h\alpha = f$; f is in x and $f = h\alpha$ for some h; f is in x and, if $\alpha(i) = \alpha(k)$, then $f(i) = f(k)$; f is in x and $f(i) = f(k)$ for each $\langle i, k\rangle \in \alpha\check{}\alpha$; f is in $(e\alpha\check{}\alpha)x$.

For any $\mathcal{S}$, we now define:

(11) $\mathcal{C}(\mathcal{S}) = \{-(\text{ran }\alpha_0) + \cdots + -(\text{ran }\alpha_n) : n \in \omega$ and $\alpha_0, \cdots, \alpha_n$ are in $\text{cl}\mathcal{S}\}$.

(12) $\mathcal{E}(\mathcal{S}) = \{\text{eqv}(\alpha_0\check{}\alpha_0 + \cdots + \alpha_n\check{}\alpha_n) : n \in \omega$ and $\alpha_0, \cdots, \alpha_n$ are in $\text{cl}\mathcal{S}\}$.

Henceforth, unless explicitly stated otherwise, the transformational operations shall be those resulting from the $\mathcal{S}$ under discussion and the extra-Boolean operations those resulting from $\langle \mathcal{S}, \mathcal{C}(\mathcal{S}), \mathcal{E}(\mathcal{S})\rangle$.

The following three identities follow easily from (8) or from (1) and (8) respectively.

(13) $(t\beta)(t\alpha) = (t\alpha\beta)$.

(14) $(t\alpha)(s\alpha)(t\alpha) = (t\alpha)$.

(15) $(s\text{ id})(t\alpha) = (t\alpha)(s\text{ id}) = (t\alpha)$.

From (11), (6), (9), (5), and (13) and from (12), (7), (10), (5), and (13) there follows the first of the two theorems now to be given.

THEOREM 1: *For any $\mathcal{S}$, each extra-Boolean operation is transformational.*[1]

---

[1] Several years ago, Charles M. Howard in [6] systematically utilized a closely related observation. This was partly done on my suggestion, which in turn was partly caused by his use of some operations $(t\alpha) = \hat{\alpha}^*$ in an earlier draft of [6]. His interesting results encouraged me to do further work in this area. Howard starts with functions $\hat{\alpha}(f) = f \circ \alpha$ such that α belongs to a certain subset $\mathcal{S}'$ of $\bigcup\{^J I : J \subseteq I\}$ and f to a certain subset $\mathcal{I}'$ of $\bigcup\{^J U : J \subseteq I\}$. In addition to $\hat{\alpha}^*$, $\hat{\alpha}^{-1}$, and certain Boolean operations, he also takes the set $\{\emptyset\}$ consisting of the empty sequence as primitive. In general, $\{\emptyset\}$ is not definable by means of his other primitives.

THEOREM 2: *If* $\mathcal{S} \subseteq \mathrm{cl}(\mathcal{I}(\mathcal{S}) + \mathcal{R}(\mathcal{S}) + \mathcal{L}(\mathcal{S}))$, *then each transformational operation is extra-Boolean.*

Proof: By $\mathcal{S} \subseteq \mathrm{cl}(\mathcal{I}(\mathcal{S}) + \mathcal{R}(\mathcal{S}) + \mathcal{L}(\mathcal{S}))$, (13), and induction, it suffices to show that $(t\alpha)$ is extra-Boolean when α is in $\mathcal{I}(\mathcal{S})$, $\mathcal{R}(\mathcal{S})$, or $\mathcal{L}(\mathcal{S})$. First assume that α is in $\mathcal{I}(\mathcal{S})$. Then $(t\alpha) = (t\alpha)(s\alpha)(t\alpha) = (t\alpha)(s\alpha\alpha)(t\alpha) = (t\alpha)(s\alpha)(s\alpha)(t\alpha) = (e\alpha\breve{}\alpha)(c - \mathrm{ran}\,\alpha)$ by (14), the idempotency of α, (5), and (10) and (9) respectively. Now assume that α is in $\mathcal{R}(\mathcal{S})$. Then α has a left inverse β in $\mathcal{L}(\mathcal{S})$. Then $(t\alpha) = (s\,\mathrm{id})(t\alpha) = (s\beta\alpha)(t\alpha) = (s\beta)(s\alpha)(t\alpha) = (s\beta)(c - \mathrm{ran}\,\alpha)$ by (15), $\beta\alpha = \mathrm{id}$, (5), and (9) respectively.[2] Finally assume that α is in $\mathcal{L}(\mathcal{S})$. Then α has a right inverse in $\mathcal{R}(\mathcal{S})$. Then $(t\alpha) = (t\alpha)(s\,\mathrm{id}) = (t\alpha)(s\alpha\beta) = (t\alpha)(s\alpha)(s\beta) = (e\alpha\breve{}\alpha)(s\beta)$ by (15), $\alpha\beta = \mathrm{id}$, (5), and (10) respectively. It follows from (5), (11), and (12) that $(t\alpha)$ is extra-Boolean in each of these cases. ♦

Let us consider a further language from the literature. It was introduced by Copeland [2] and has been investigated, quite recently, by Demaree [3_D]. In [2], Copeland lets I be the set of all integers (positive, 0, negative) and takes as primitive extra-Boolean operations exactly those (cJ), (eM), and $(s\alpha)$ such that $J = \{0\}$, $M = \{\langle 0, 1\rangle\}$, and $\alpha \in \{(+1), (-1), (0, 1)\}$ where $(+1)(i) = i + 1$ and $(-1)(i) = i - 1$ for each $i \in I$. In effect, he notes that each (i, k) is in $\mathrm{cl}\{(+1), (-1), (0, 1)\}$ and hence that the extra-Boolean operations[3] expressible in his language, i.e., those resulting from $\langle\{(+1), (-1), (0, 1)\}, \{\{0\}\}, \{\{\langle 0, 1\rangle\}\}\rangle$, also result from $\langle \mathrm{cl}\{(0/1), (+1), (-1), (0, 1)\}, \mathcal{C}_\omega, \mathcal{E}_\omega\rangle$.

We now apply the last two theorems to some languages.[4]

---

[2] This identity, and the fact that in this case $(t\alpha)$ is extra-Boolean, were first noticed by Richard Thompson.

[3] The notion of an extra-Boolean operation was partly suggested by [2]. We shall sometimes assume tacitly that $0 \in I$, so that $\{(+1), (-1)\} \subseteq \mathcal{S}$, in conjunction with our convention that $\mathcal{S} \subseteq {}^I I$, implies that I contains all integers.

[4] For another result of this kind see Theorem 7.6 of [3]. For results when not all sequences are of the same length, see Theorems 7.11 and 7.12 of [3] and the remarks following their proofs.

THEOREM 3:

(a) *The extra-Boolean operations expressible in the language of cylindric algebras are the transformational operations resulting from* $\mathrm{cl}\{\alpha \in \mathcal{S}_\omega : \alpha\alpha = \alpha\}$.

(b) *For any infinite cardinal* κ, *the extra-Boolean operations expressible in Lucas' language are the transformational operations resulting from* $\mathcal{S}_\kappa$.

(c) *The extra-Boolean operations expressible in Copeland's language are the transformational operations resulting from* $\mathrm{cl}\{(0/1), (+1), (-1), (0, 1)\}$.

Proof:

(a) Let $\mathcal{S} = \mathrm{cl}\{\alpha \in \mathcal{S}_\omega : \alpha\alpha = \alpha\}$. We saw earlier that the extra-Boolean operations expressible in the language of cylindric algebras also result from $\langle \mathcal{S}, \mathcal{C}_\omega, \mathcal{E}_\omega \rangle$. Now $\mathcal{C}_\omega = \mathcal{C}(\mathcal{S})$ and $\mathcal{E}_\omega = \mathcal{E}(\mathcal{S})$, as is easily verified. Also $\mathcal{S} = \mathrm{cl}\mathcal{I}(\mathcal{S})$ and hence $\mathcal{S} \subseteq \mathrm{cl}(\mathcal{I}(\mathcal{S}) + \mathcal{R}(\mathcal{S}) + \mathcal{L}(\mathcal{S}))$. Hence, by Theorems 1 and 2, the extra-Boolean operations resulting from $\langle \mathcal{S}, \mathcal{C}(\mathcal{S}), \mathcal{E}(\mathcal{S}) \rangle$ are the transformational operations resulting from $\mathcal{S}$.

(b) Let κ be any infinite cardinal, and let $\mathcal{S} = \mathcal{S}_\kappa$. One sees easily that $\mathcal{C}_\kappa = \mathcal{C}(\mathcal{S})$ and $\mathcal{E}_\kappa = \mathcal{E}(\mathcal{S})$. Hence the extra-Boolean operations expressible in Lucas' language result from $\langle \mathcal{S}, \mathcal{C}(\mathcal{S}), \mathcal{E}(\mathcal{S}) \rangle$. If I is finite or $\kappa = \omega$, then $\mathcal{S}$ is $\{\beta\gamma : \beta \in \mathcal{R}(\mathcal{S}) \cdot \mathcal{L}(\mathcal{S})$ and $\gamma \in \mathcal{I}(\mathcal{S})\}$, as we noted earlier. If I is infinite and $\kappa > \omega$, then $\mathcal{S}$ is $\{\beta\gamma : \beta \in \mathcal{L}(\mathcal{S})$ and $\gamma \in \mathcal{R}(\mathcal{S})\}$, as is also not hard to show. Hence, in either case, $\mathcal{S} \subseteq \mathrm{cl}(\mathcal{I}(\mathcal{S}) + \mathcal{R}(\mathcal{S}) + \mathcal{L}(\mathcal{S}))$. Hence, by Theorems 1 and 2, the extra-Boolean operations resulting from $\langle \mathcal{S}, \mathcal{C}(\mathcal{S}), \mathcal{E}(\mathcal{S}) \rangle$ are the transformational operations resulting from $\mathcal{S}$.

(c) Let $\mathcal{S} = \mathrm{cl}\{(0/1), (+1), (-1), (0, 1)\}$. By induction one shows that $\mathcal{S}$ is $\{\beta\gamma : \beta \in \mathrm{cl}\{(+1), (-1)\}$ and $\gamma \in \mathcal{S}_\omega\}$. Also, as Copeland indicated, the extra-Boolean operations expressible in his language result from $\langle \mathcal{S}, \mathcal{C}_\omega, \mathcal{E}_\omega \rangle$. The rest of the proof is similar to those of (a) and (b). ♦

The situation is different for the extra-Boolean operations expressible in Halmos' language of polyadic algebras with diagonal elements. As was essentially noted earlier, they result from

$\langle {}^{I}I,\ \{J : J \subseteq I\},\ \{\{\langle i, k\rangle\} : \langle i, k\rangle \in {}^{2}I\}\rangle$. For finite I, they also result from $\langle \mathcal{S}_\omega, \mathcal{C}_\omega, \mathcal{E}_\omega\rangle$ and hence are the transformational operations resulting from $\mathcal{S}_\omega$. However, if I is infinite, then for each $\mathcal{S}$ either $\{J : J \subseteq I\}$ is bigger than $\mathcal{C}(\mathcal{S})$ or $\{\{\langle i, k\rangle\} : \langle i, k\rangle \in {}^{2}I\}$ is smaller than $\mathcal{E}(\mathcal{S})$. Nevertheless, for $\kappa > \| I \|$, the extra-Boolean operations expressible in Halmos' language form a subset of the transformational operations resulting from $\mathcal{S}_\kappa$.

For some languages, there seems to exist no $\langle \mathcal{S}, \mathcal{C}, \mathcal{E}\rangle$ such that the operations resulting from $\langle \mathcal{S}, \mathcal{C}, \mathcal{E}\rangle$ and the Boolean operations generate by substitution the operations of the language. Among these seem to be the language of Quine [11], where the arguments x of operations are sets of sequences of a finite length $l(x)$ which depends on x, that of Monk [10], where the arguments x are sets of sequences of arbitrary finite length, that of Howard [6], where the arguments x are sets of functions f whose domain $J \subseteq I$ depends on f, and that of the classical calculus of binary relations, where $I = 2$ and among the operations is the relative product.

Addendum: Just prior to submission of this paper, Richard Thompson observed: If $\alpha\beta\alpha = \alpha$, then $(t\alpha) = (e\alpha\check{}\alpha)(s\beta)(c - \operatorname{ran}\alpha)$. For proof, one uses (10) and (9), (5), $\alpha\beta\alpha = \alpha$, and (14). Now let α be a *regular* element of cl$\mathcal{S}$ if and only if there is some β in cl$\mathcal{S}$ such that $\alpha\beta\alpha = \alpha$ (cf. [1]). It follows, as he pointed out, that in Theorem 2 one can replace the condition $\mathcal{S} \subseteq \operatorname{cl}(\mathcal{I}(\mathcal{S}) + \mathcal{R}(\mathcal{S}) + \mathcal{L}(\mathcal{S}))$ by the weaker condition that $\mathcal{S} \subseteq \operatorname{cl}\{\alpha : \alpha$ is a regular element of cl$\mathcal{S}\}$. One can also apply his observation to §4 below to replace D4, $\cdots$, D7, strengthen Lemma 27, Theorem 28, and Theorem 29, and simplify their proofs.

2. CONDITIONS ON ANTIREPRESENTATIONS OF SEMIGROUPS

The region of set theory considered in §1 can be divided into two strata: A lower stratum of mappings $\hat{\alpha}$ transforming sequences into sequences, and an upper stratum of operations on sets of sequences, including Boolean operations and operations $\hat{\alpha}^{-1}$ and $\hat{\alpha}^*$. In §2, we shall abstract progressively from the lower stratum.

An *(abstract) semigroup* shall be any $\langle \mathcal{S}, \circ\rangle$ such that $\mathcal{S}$ is a set,

$\circ$ is a 2-ary operation on $\mathcal{S}$, and $\circ$ is associative, i.e., any elements α, β, γ of $\mathcal{S}$ satisfy $(\alpha \circ \beta) \circ \gamma = \alpha \circ (\beta \circ \gamma)$. In particular, each semigroup of transformations is a semigroup. We shall often identify $\langle \mathcal{S}, \circ \rangle$ with $\mathcal{S}$. A *prerepresentation of* a semigroup $\mathcal{S}$ shall be any function whose domain is $\mathcal{S}$ and whose values are, for some set W, transformations on W. It shall be *trivial* if and only if each of its values is the identity transformation on W.

In §2, $\mathcal{S}$ shall be a semigroup and $\hat{}$ a prerepresentation of $\mathcal{S}$, both of which will remain fixed. The range of $\hat{}$ shall be $\mathfrak{J}$.

Henceforth, $\alpha, \beta, \gamma, \cdots$ shall be elements of $\mathcal{S}$ (instead of, as in §1, transformations on I). We let $\alpha\beta = \alpha \circ \beta$. If $\mathcal{S}$ has an identity element, i.e., an element γ such that $\gamma\alpha = \alpha = \alpha\gamma$ for every α, then we let id be this, necessarily unique, element. Note that id differs from the identity transformation on I, and hence from the id of §1, when, for example, $\mathcal{S}$ is $\{\gamma\}$ and γ is an idempotent transformation on I with ran $\gamma \neq I$.

Assume, for a moment, that $\mathcal{S}$ consists of transformations on I. If both $\alpha(i) = i$ and $\alpha(j) \neq i$ for each $j \neq i$, then i shall be an *isolated point for* α. Thus, i is an isolated point for α if and only if $\alpha^{-1}\{i\} = \{i\}$. We let isl$\mathcal{S}$ be the set of those i which are an isolated point for each α in $\mathcal{S}$. We let $\mathcal{S}$ be *trivial* if and only if each α in $\mathcal{S}$ is the identity transformation on I. It follows that $\mathcal{S}$ is trivial if and only if isl$\mathcal{S} = I$.

We now list six conditions on $\hat{}$. Among these, $T_t3, \cdots, T_t6$ are applicable only when $\mathcal{S}$ is a semigroup of transformations (on some V). Evidently, if $\hat{}$ satisfies T1, then $\mathfrak{J}$ is closed under composition and hence a semigroup of transformations. In that case we call $\hat{}$, and also $\mathfrak{J}$, an *antirepresentation* of $\mathcal{S}$. If T2 is satisfied, then we shall say that $\mathfrak{J}$ *has* (or $\hat{}$ *yields*) *commuting equivalences.*

T1. $\widehat{\alpha\beta} = \hat{\beta}\hat{\alpha}$.

T2. $\hat{\alpha}^{\smallsmile}\hat{\alpha}\hat{\beta}^{\smallsmile}\hat{\beta} = \hat{\beta}^{\smallsmile}\hat{\beta}\hat{\alpha}^{\smallsmile}\hat{\alpha}$.

T_t3. If $\text{eqv}(\alpha^{\smallsmile}\alpha + \beta^{\smallsmile}\beta) = \gamma^{\smallsmile}\gamma$, then $\text{ran}\,\hat{\alpha} \cdot \text{ran}\,\hat{\beta} = \text{ran}\,\hat{\gamma}$.

T_t4. If $(\beta^{\smallsmile}\beta) \upharpoonright (\text{ran}\,\alpha) = \beta^{\smallsmile}\beta$, then $\hat{\alpha}^{-1}(\hat{\alpha}^*(\text{ran}\,\hat{\beta})) = \text{ran}\,\hat{\beta}$.

T_t5. If $(\beta^{\smallsmile}\beta) \upharpoonright (\text{ran}\,\alpha) \neq \beta^{\smallsmile}\beta$, then $-\hat{\alpha}^{-1}(\hat{\alpha}^*(-\text{ran}\,\hat{\beta})) = \text{isl}\mathfrak{J}$.

T_t6. If $\hat{}$ is nontrivial, then $\alpha^{\smallsmile}\alpha \subseteq \beta^{\smallsmile}\beta$ if and only if $\text{ran}\,\hat{\beta} \subseteq \text{ran}\,\hat{\alpha}$.

THEOREM 4: *Assume that $\mathcal{S}$ is a semigroup of transformations on I and that $\hat{}$ is the restriction of the $\hat{}$ of §1 to $\mathcal{S}$. Then $\hat{}$ satisfies* T1, T2, $T_t3, \cdots, T_t6$.

Proof: For any α, β, and f one has $\hat{\beta}(\hat{\alpha}(f)) = \hat{\beta}(f\alpha) = (f\alpha)\beta = f(\alpha\beta) = \widehat{\alpha\beta}(f)$. Hence $\hat{}$ satisfies T1.

Next, note that each of the two conditions $\langle f, h\rangle \in \hat{\alpha}^{\smile}\hat{\alpha}\,\hat{\beta}^{\smile}\hat{\beta}$, $\langle f, h\rangle \in \hat{\beta}^{\smile}\hat{\beta}\,\hat{\alpha}^{\smile}\hat{\alpha}$ is equivalent to the condition that $f(i) = h(i)$ for every i in $\operatorname{ran}\alpha \cdot \operatorname{ran}\beta$. It follows that $\hat{}$ satisfies T2.

For the rest of the proof it is important to note that $f \in \operatorname{ran}\hat{\delta}$ is equivalent to the condition that $f = g\delta$ for some g and hence to the condition that $\delta^{\smile}\delta \subseteq f^{\smile}f$.

Now assume that $\operatorname{eqv}(\alpha^{\smile}\alpha + \beta^{\smile}\beta) = \gamma^{\smile}\gamma$. It follows that $\operatorname{ran}\hat{\gamma} \subseteq \operatorname{ran}\hat{\alpha} \cdot \operatorname{ran}\hat{\beta}$. Also, if $f \in \operatorname{ran}\hat{\alpha} \cdot \operatorname{ran}\hat{\beta}$, then $\alpha^{\smile}\alpha + \beta^{\smile}\beta \subseteq f^{\smile}f$ and hence $\gamma^{\smile}\gamma = \operatorname{eqv}(\alpha^{\smile}\alpha + \beta^{\smile}\beta) \subseteq \operatorname{eqv}(f^{\smile}f) = f^{\smile}f$. Hence $\operatorname{ran}\hat{\alpha} \cdot \operatorname{ran}\hat{\beta} \subseteq \operatorname{ran}\hat{\gamma}$. It follows that $\hat{}$ satisfies T_t3.

Clearly, $\operatorname{ran}\hat{\beta} \subseteq \hat{\alpha}^{-1}(\hat{\alpha}^*(\operatorname{ran}\hat{\beta}))$. Now assume that $(\beta^{\smile}\beta) \upharpoonright (\operatorname{ran}\alpha) = \beta^{\smile}\beta$. Let $f \in \hat{\alpha}^{-1}(\hat{\alpha}^*(\operatorname{ran}\hat{\beta}))$. Then there is some $h \in \operatorname{ran}\hat{\beta}$ such that f and h agree on $\operatorname{ran}\alpha$. Now consider any $\langle j, k\rangle \in \beta^{\smile}\beta$. If $j = k$, then $f(j) = f(k)$. Now suppose that $j \neq k$. Since $(\beta^{\smile}\beta) \upharpoonright (\operatorname{ran}\alpha) = \beta^{\smile}\beta$, therefore $j \in \operatorname{ran}\alpha$ and $k \in \operatorname{ran}\alpha$. Since $h \in \operatorname{ran}\hat{\beta}$ and hence $\beta^{\smile}\beta \subseteq h^{\smile}h$, therefore $h(j) = h(k)$. Since f and h agree on $\operatorname{ran}\alpha$, therefore again $f(j) = f(k)$. Thus $\beta^{\smile}\beta \subseteq f^{\smile}f$ and hence $f \in \operatorname{ran}\hat{\beta}$. Hence $\hat{\alpha}^{-1}(\hat{\alpha}^*(\operatorname{ran}\hat{\beta})) \subseteq \operatorname{ran}\hat{\beta}$. It follows that $\hat{}$ satisfies T_t4.

Let $f \in \operatorname{isl}\mathfrak{J}$. Then $\hat{\alpha}(f) = f = \hat{\beta}(f)$ and, for any h, if $h \neq f$ then $\hat{\alpha}(h) \neq f \neq \hat{\beta}(h)$. It follows that $f \notin -\operatorname{ran}\hat{\beta}$, hence $f \notin \hat{\alpha}^*(-\operatorname{ran}\hat{\beta})$, hence $f \notin \hat{\alpha}^{-1}(\hat{\alpha}^*(-\operatorname{ran}\hat{\beta}))$, and hence $f \in -\hat{\alpha}^{-1}(\hat{\alpha}^*(-\operatorname{ran}\hat{\beta}))$. Therefore $\operatorname{isl}\mathfrak{J} \subseteq -\hat{\alpha}^{-1}(\hat{\alpha}^*(-\operatorname{ran}\hat{\beta}))$.

Assume, first, that $\|U\| \leqq 1$. Then $-\hat{\alpha}^{-1}(\hat{\alpha}^*(-\operatorname{ran}\hat{\beta})) \subseteq {}^IU = \operatorname{isl}\mathfrak{J}$. Assume now that $\|U\| > 1$ and that $(\beta^{\smile}\beta) \upharpoonright (\operatorname{ran}\alpha) \neq \beta^{\smile}\beta$. Then there is some $j \in -\operatorname{ran}\alpha$ and some $k \neq j$ such that $\beta(j) = \beta(k)$. Consider any f. Since $\|U\| > 1$, there is some h such that $h(i) = f(i)$ for each $i \neq j$ and such that $h(j) \neq f(k) = h(k)$. Then $\beta^{\smile}\beta \nsubseteq h^{\smile}h$ and hence $h \in -\operatorname{ran}\hat{\beta}$. Also f and h agree on $\operatorname{ran}\alpha$ and hence $f \in \hat{\alpha}^{-1}(\hat{\alpha}^*\{h\}) \subseteq \hat{\alpha}^{-1}(\hat{\alpha}^*(-\operatorname{ran}\hat{\beta}))$. It follows that $-\hat{\alpha}^{-1}(\hat{\alpha}^*(-\operatorname{ran}\hat{\beta})) = \varnothing \subseteq \operatorname{isl}\mathfrak{J}$. Thus $\hat{}$ satisfies T_t5.

Assume that $\alpha^{\smile}\alpha \subseteq \beta^{\smile}\beta$. Consider any $f \in \operatorname{ran}\hat{\beta}$. Then $\beta^{\smile}\beta \subseteq$

$f^{\smile}f$. Hence $\alpha^{\smile}\alpha \subseteq f^{\smile}f$, and therefore $f \in \operatorname{ran} \hat{\alpha}$. Therefore $\alpha^{\smile}\alpha \subseteq \beta^{\smile}\beta$ implies that $\operatorname{ran} \hat{\beta} \subseteq \operatorname{ran} \hat{\alpha}$.

Now assume that $\hat{}$ and hence $\mathfrak{J}$ is nontrivial. Then $\| U \| > 1$. Suppose that $\alpha^{\smile}\alpha \nsubseteq \beta^{\smile}\beta$. Then some $\langle j, k\rangle$ is in $\alpha^{\smile}\alpha$ and not in $\beta^{\smile}\beta$. Since $\| U \| > 1$, there is some h such that $h(j) \neq h(k)$ and such that $h(i) = h(k)$ for each $i \neq j$. Then $\alpha^{\smile}\alpha \nsubseteq h^{\smile}h$ and $\beta^{\smile}\beta \subseteq h^{\smile}h$. Hence $h \notin \operatorname{ran} \hat{\alpha}$ and $h \in \operatorname{ran} \hat{\beta}$, so that $\operatorname{ran} \hat{\beta} \nsubseteq \operatorname{ran} \hat{\alpha}$. Hence, if $\operatorname{ran} \hat{\beta} \subseteq \operatorname{ran} \hat{\alpha}$, then $\alpha^{\smile}\alpha \subseteq \beta^{\smile}\beta$. Thus $\hat{}$ satisfies T_t6. ♦

We let $\alpha \leqq_R \beta$, or let α be a *right divisor of* β, if and only if there is some γ such that $\gamma\alpha = \beta$. We let $\alpha \leqq_L \beta$, or let α be a *left divisor of* β, if and only if there is some γ such that $\alpha\gamma = \beta$.

Assume that $\alpha \leqq_R \beta$ and $\beta \leqq_R \gamma$. Then there are α' and β' and that $\alpha'\alpha = \beta$ and $\beta'\beta = \gamma$. Then $\gamma = \beta'\beta = \beta'\alpha'\alpha$. Hence $\alpha \leqq_R \gamma$. It follows that $\leqq_R$ is transitive. Similarly $\leqq_L$ is transitive. If $\mathcal{S}$ has an identity, then $\leqq_R$ and $\leqq_L$ are also reflexive. In that case, $\leqq_R$ and $\leqq_L$ shall be a *partial preordering* of $\mathcal{S}$.

Let $\mathcal{S}'$ be a subset of $\mathcal{S}$. If $\alpha \leqq_R \beta$ for every β in $\mathcal{S}'$, then α shall be an $\leqq_R$ *lower bound of* $\mathcal{S}'$. If $\beta \leqq_R \alpha$ for every β in $\mathcal{S}'$, then α shall be an $\leqq_R$ *upper bound of* $\mathcal{S}'$. If α is an $\leqq_R$ lower bound of $\mathcal{S}'$ and also an $\leqq_R$ upper bound of the set of all $\leqq_R$ lower bounds of $\mathcal{S}'$, then α shall be an $\leqq_R$ *greatest lower bound*, or $\leqq_R$ *glb, of* $\mathcal{S}'$. If α is an $\leqq_R$ upper bound of $\mathcal{S}'$ and also an $\leqq_R$ lower bound of the set of all $\leqq_R$ upper bounds of $\mathcal{S}'$, then α shall be an $\leqq_R$ *least upper bound*, or $\leqq_R$ *lub, of* $\mathcal{S}'$. We let $\alpha =_R \beta$ if and only if $\alpha \leqq_R \beta$ and $\beta \leqq_R \alpha$. Note that $\alpha =_R \beta$ does not imply that $\alpha = \beta$. We let $\alpha \nleqq_R \beta$ if and only if $\alpha \leqq_R \beta$ does not hold. For $\leqq_L$, the corresponding notions shall be defined similarly. Note that $\mathcal{S}'$ may have more than one $\leqq_R$ glb, and likewise more than one $\leqq_R$ lub, $\leqq_L$ glb, or $\leqq_L$ lub.

Next, we list certain conditions on a semigroup of transformations. Each of these is a condition that a certain set-theoretic requirement concerning elements of the semigroup be equivalent to an algebraic requirement. If a semigroup of transformations satisfies conditions E_L2, E_L3, and E_L4 below then we shall say that it *has suitable ranges*. For convenience, we shall formulate these conditions as conditions on $\mathcal{S}$, assuming for this purpose that $\mathcal{S}$ is a semigroup of transformations.

E_R1. $\breve{\alpha}\alpha \subseteq \breve{\beta}\beta$ if and only if $\alpha \leqq_R \beta$.
E_L1. ran $\alpha \subseteq$ ran β if and only if $\beta \leqq_L \alpha$.
E_R2. $\text{eqv}(\breve{\alpha}\alpha + \breve{\beta}\beta) = \breve{\gamma}\gamma$ if and only if γ is an $\leqq_R$ lub of $\{\alpha, \beta\}$.
E_L2. ran $\alpha \cdot$ ran $\beta =$ ran γ if and only if γ is an $\leqq_L$ lub of $\{\alpha, \beta\}$.
E_R3. $(\breve{\beta}\beta) \upharpoonright (\text{ran}\,\alpha) = \breve{\beta}\beta$ if and only if β is an $\leqq_R$ lower bound of $\{\delta: \beta\alpha \leqq_R \delta\alpha\}$.
E_L3. $\alpha^{-1}(\alpha^*(\text{ran}\,\beta)) = \text{ran}\,\beta$ if and only if β is an $\leqq_L$ lower bound of $\{\delta: \alpha\beta \leqq_L \alpha\delta\}$.
E_L4. $-\alpha^{-1}(\alpha^*(-\text{ran}\,\beta)) \neq \text{isl}\mathcal{S}$ if and only if β is an $\leqq_L$ lower bound of $\{\delta: \alpha\beta \leqq_L \alpha\delta\}$.
E_L5. ran $\alpha \cdot$ ran $\beta \neq \varnothing$ if and only if $\{\alpha, \beta\}$ has an $\leqq_L$ upper bound.

Among the following conditions on a semigroup, different ones will be used for different purposes, and S6, $\cdots$, S9 will not be used until §3. Those which are applicable only to semigroups of transformations are marked accordingly. Any semigroup satisfying S1 shall be *regular* (cf. [1]). For convenience we shall again assume for a moment that $\mathcal{S}$ is a semigroup of transformations, and formulate these conditions as conditions on $\mathcal{S}$. As before, α and β shall be arbitrary elements of $\mathcal{S}$.

S0. $\mathcal{S} = \text{cl}(\mathcal{I}(\mathcal{S}) + \mathcal{R}(\mathcal{S}) + \mathcal{L}(\mathcal{S}))$.
S1. There is some γ such that $\alpha\gamma\alpha = \alpha$.
S2. There is some γ such that γ is an $\leqq_R$ lub of $\{\alpha, \beta\}$.
S_t2. There is some γ such that $\breve{\gamma}\gamma = \text{eqv}(\breve{\alpha}\alpha + \breve{\beta}\beta)$.
S_t3. Each replacement on dom α is in $\mathcal{S}$.
S_t4. If $\breve{\alpha}\alpha = {}^2(\text{dom}\,\alpha)$, then for each k in dom α there is some γ such that $\breve{\gamma}\gamma = (\breve{\alpha}\alpha) \upharpoonright -\{k\}$.
S5. If α is idempotent, β is idempotent, $\alpha \leqq_R \beta$, and $\alpha \leqq_L \beta$, then there is some γ such that γ is idempotent, $\gamma\alpha = \beta$, and γ is an $\leqq_R$ lower bound of $\{\delta: \gamma\alpha \leqq_R \delta\alpha\}$.
S6. If β is idempotent, $\alpha \leqq_L \beta\alpha$, and id is an $\leqq_L$ glb of $\{\alpha, \beta\}$, then β is an $\leqq_R$ lower bound of $\{\delta: \beta\alpha \leqq_R \delta\alpha\}$.
S7. There are α' and β' such that $\alpha' \leqq_L \alpha$, $\beta' \leqq_L \beta$, $\{\alpha', \beta'\}$ and $\{\alpha, \beta\}$ have the same $\leqq_L$ upper bounds, and id is an $\leqq_L$ glb of $\{\alpha', \beta'\}$.

S8. If $\{\alpha, \beta\}$ has an $\leqq_L$ upper bound, then there are α' and β' such that α' is idempotent, β' is idempotent, $\alpha' =_L \alpha$, $\beta' =_L \beta$, and $\alpha'\beta' = \beta'\alpha'$.

S9. There is some γ such that γ is an $\leqq_R$ lower bound of $\{\delta: \gamma\alpha \leqq_R \delta\alpha\}$ and $\gamma\alpha$ is an $\leqq_R$ lub of $\{\alpha, \beta\}$.

Some elementary facts about semigroups will be useful. Henceforth, we shall often omit easy proofs.

LEMMA 5: *For parts* (c) *and* (d) *assume that* $\mathcal{S}$ *is regular.*

(a)$_R$ *If* α *is idempotent and* $\alpha \leqq_R \beta$, *then* $\beta\alpha = \beta$.

(a)$_L$ *If* α *is idempotent and* $\alpha \leqq_L \beta$, *then* $\alpha\beta = \beta$.

(b)$_R$ *If* α *is idempotent,* β *is idempotent, and* $\alpha \leqq_R \beta$, *then* $\alpha\beta$ *is idempotent and* $\alpha\beta =_R \beta$.

(b)$_L$ *If* α *is idempotent,* β *is idempotent, and* $\alpha \leqq_L \beta$, *then* $\beta\alpha$ *is idempotent and* $\beta\alpha =_L \beta$.

(c)$_R$ *There is some* γ *such that* γ *is idempotent and* $\gamma =_R \alpha$.

(c)$_L$ *There is some* γ *such that* γ *is idempotent and* $\gamma =_L \alpha$.

(d)$_R$ *If* $\alpha \leqq_R \beta$, *then there are* α' *and* β' *such that* α' *is idempotent,* β' *is idempotent,* $\alpha =_R \alpha'$, $\beta =_R \beta'$, *and* $\alpha' \leqq_L \beta'$.

(d)$_L$ *If* $\alpha \leqq_L \beta$, *then there are* α' *and* β' *such that* α' *is idempotent,* β' *is idempotent,* $\alpha =_L \alpha'$, $\beta =_L \beta'$, *and* $\alpha' \leqq_R \beta'$.

Proof:

(b)$_R$ Clearly, $\beta \leqq_R \alpha\beta$. By 5(a)$_R$, $\beta\alpha\beta = \beta\beta = \beta$. It follows that $\alpha\beta \leqq_R \beta$ and that $\alpha\beta\alpha\beta = \alpha\beta$.

(c)$_R$ We use the argument of [1], p. 26. By the regularity of $\mathcal{S}$, there is some β such that $\alpha\beta\alpha = \alpha$. Let $\gamma = \beta\alpha$. Then $\alpha \leqq_R \gamma$ and $\gamma \leqq_R \alpha\beta\alpha = \alpha$. Also $\gamma\gamma = \beta\alpha\beta\alpha = \beta\alpha = \gamma$.

(d)$_R$ Assume that $\alpha \leqq_R \beta$. By 5(c)$_R$, there are α' and β'' such that $\alpha'\alpha' = \alpha'$, $\beta''\beta'' = \beta''$, $\alpha' =_R \alpha$, and $\beta'' =_R \beta$. Let $\beta' = \alpha'\beta''$. Then $\alpha' \leqq_L \beta'$. Also $\alpha' \leqq_R \beta''$ so that, by 5(b)$_R$, β' is idempotent and $\beta' =_R \beta'' =_R \beta$. ♦

Most parts of the following lemma seem to be well known. Part (e) will not be used, but has been added to round out the picture.

LEMMA 6: *Assume that* $\mathcal{S}$ *is a regular semigroup of transformations.*

(a) $\mathcal{S}$ *satisfies* E_R1.
(b) $\mathcal{S}$ *satisfies* E_L1.
(c) *If* $\mathcal{S}$ *satisfies* S_t2, *then* $\mathcal{S}$ *satisfies* E_R2.
(d) *If* $\mathcal{S}$ *satisfies* S_t3 *and* S_t4, *then* $\mathcal{S}$ *satisfies* E_R3.
(e) *If* $\mathcal{S}$ *satisfies* S_t3, *then* $\mathcal{S}$ *satisfies* E_L3.

Proof: For convenience, let $\mathcal{S}$ be a set of transformations on I.

(a) First assume that $\alpha \leqq_R \beta$. Then $\gamma\alpha = \beta$ for some γ. Then $\alpha(j) = \alpha(k)$ implies that $\beta(j) = \gamma\alpha(j) = \gamma\alpha(k) = \beta(k)$. Hence $\alpha^{\smile}\alpha \subseteq \beta^{\smile}\beta$. Now assume that $\mathcal{S}$ is regular and that $\alpha^{\smile}\alpha \subseteq \beta^{\smile}\beta$. There is some δ such that $\alpha\delta\alpha = \alpha$. Let $\gamma = \beta\delta$. Consider any $i \in I$. Since $\alpha\delta\alpha(i) = \alpha(i)$, therefore $\langle \delta\alpha(i), i\rangle \in \alpha^{\smile}\alpha \subseteq \beta^{\smile}\beta$ and hence $\gamma\alpha(i) = \beta\delta\alpha(i) = \beta(i)$. Therefore $\gamma\alpha = \beta$ and $\alpha \leqq_R \beta$.
(b) First assume that $\beta \leqq_L \alpha$. Then $\beta\gamma = \alpha$ for some γ. Hence $\operatorname{ran} \alpha \subseteq \operatorname{ran} \beta$. Now assume that $\mathcal{S}$ is regular and that $\operatorname{ran} \alpha \subseteq \operatorname{ran} \beta$. There is some δ such that $\beta\delta\beta = \beta$. Let $\gamma = \delta\alpha$. Consider any $i \in I$. There is some $k \in I$ such that $\alpha(i) = \beta(k)$. Then $\beta\gamma(i) = \beta\delta\alpha(i) = \beta\delta\beta(k) = \beta(k) = \alpha(i)$. Therefore $\beta\gamma = \alpha$ and $\beta \leqq_L \alpha$.
(c) Assume that $\mathcal{S}$ is regular. First assume that $\operatorname{eqv}(\alpha^{\smile}\alpha+\beta^{\smile}\beta) = \gamma^{\smile}\gamma$. Then $\alpha^{\smile}\alpha \subseteq \gamma^{\smile}\gamma$ and $\beta^{\smile}\beta \subseteq \gamma^{\smile}\gamma$. By 6(a), $\alpha \leqq_R \gamma$ and $\beta \leqq_R \gamma$. Now suppose that $\alpha \leqq_R \delta$ and $\beta \leqq_R \delta$. Then $\alpha^{\smile}\alpha \subseteq \delta^{\smile}\delta$ and $\beta^{\smile}\beta \subseteq \delta^{\smile}\delta$. Then $\alpha^{\smile}\alpha + \beta^{\smile}\beta \subseteq \delta^{\smile}\delta$, and hence $\gamma^{\smile}\gamma = \operatorname{eqv}(\alpha^{\smile}\alpha + \beta^{\smile}\beta) \subseteq \operatorname{eqv}(\delta^{\smile}\delta) = \delta^{\smile}\delta$. By 6(a), $\gamma \leqq_R \delta$. Therefore γ is an $\leqq_R$ lub of $\{\alpha, \beta\}$.

Now assume that $\mathcal{S}$ satisfies S_t2 and that γ is an $\leqq_R$ lub of $\{\alpha, \beta\}$. There is some δ such that $\delta^{\smile}\delta = \operatorname{eqv}(\alpha^{\smile}\alpha + \beta^{\smile}\beta)$. As we just proved, δ is an $\leqq_R$ lub of $\{\alpha, \beta\}$. Hence $\gamma \leqq_R \delta$ and $\delta \leqq_R \gamma$. By 6(a), $\gamma^{\smile}\gamma = \delta^{\smile}\delta = \operatorname{eqv}(\alpha^{\smile}\alpha + \beta^{\smile}\beta)$.

(d) Assume that $\mathcal{S}$ is regular. First assume that $(\beta^{\smile}\beta) \restriction (\operatorname{ran} \alpha) = \beta^{\smile}\beta$. Consider any γ such that $\beta\alpha \leqq_R \gamma\alpha$. Then $(\beta\alpha)^{\smile}(\beta\alpha) \subseteq (\gamma\alpha)^{\smile}(\gamma\alpha)$. Consider any $\langle i, j\rangle \in \beta^{\smile}\beta$ such that $i \neq j$. Then $i \in \operatorname{ran} \alpha$ and $j \in \operatorname{ran} \alpha$. Hence $i = \alpha(i')$ and $j = \alpha(j')$ for some i' and j'. Then $\beta\alpha(i') = \beta(i) = \beta(j) = \beta\alpha(j')$, and hence $\langle i', j'\rangle \in$

$(\beta\alpha)\breve{}(\beta\alpha) \subseteq (\gamma\alpha)\breve{}(\gamma\alpha)$. Hence $\gamma(i) = \gamma\alpha(i') = \gamma\alpha(j') = \gamma(j)$, so that $\langle i,j\rangle \in \gamma\breve{}\gamma$. Thus $\beta\breve{}\beta \subseteq \gamma\breve{}\gamma$ and hence, by 6(a), $\beta \leqq_R \gamma$. Therefore β is an $\leqq_R$ lower bound of $\{\gamma: \beta\alpha \leqq_R \gamma\alpha\}$.

Now assume that $\mathcal{S}$ satisfies S_t3 and S_t4, and that $(\beta\breve{}\beta) \restriction \text{ran}\,\alpha \neq \beta\breve{}\beta$. Then there is some $i \in -\text{ran}\,\alpha$ and some $j \neq i$ such that $\beta(i) = \beta(j)$. Suppose first that $\beta\breve{}\beta \neq I \times I$. Then, for some k, $\beta(k) \neq \beta(i) = \beta(j)$. By S_t3, $\beta \circ (i/k)$ is in $\mathcal{S}$. Call it γ. Since $i \in -\text{ran}\,\alpha$, therefore $\beta\alpha = \gamma\alpha$ and hence $(\beta\alpha)\breve{}(\beta\alpha) = (\gamma\alpha)\breve{}(\gamma\alpha)$. By 6(a), $\beta\alpha \leqq_R \gamma\alpha$. Since $\gamma(i) = \beta(k) \neq \beta(i) = \beta(j) = \gamma(j)$, therefore $\langle i, j\rangle$ is in $\beta\breve{}\beta$ and not in $\gamma\breve{}\gamma$. Hence $\beta\breve{}\beta \nsubseteq \gamma\breve{}\gamma$ and therefore $\beta \nleqq_R \gamma$. Therefore β is not an $\leqq_R$ lower bound of $\{\gamma: \beta\alpha \leqq_R \gamma\alpha\}$. Suppose now that $\beta\breve{}\beta = I \times I$. By S_t4, there is some γ such that, for each $j \neq i$ and $k \neq i$, $\gamma(k) = \gamma(j) \neq \gamma(i)$. Since $\beta(j) = \beta(i)$, therefore $\beta\breve{}\beta \nsubseteq \gamma\breve{}\gamma$. Since $i \in -\text{ran}\,\alpha$, therefore $(\gamma\alpha)\breve{}(\gamma\alpha) = I \times I$ and hence $(\beta\alpha)\breve{}(\beta\alpha) \subseteq (\gamma\alpha)\breve{}(\gamma\alpha)$. By 6(a), $\beta\alpha \leqq_R \gamma\alpha$. Hence again β is not an $\leqq_R$ lower bound of $\{\gamma: \beta\alpha \leqq_R \gamma\alpha\}$.

(e) Assume that $\mathcal{S}$ is regular. First assume that $\alpha^{-1}(\alpha^*(\text{ran}\,\beta)) = \text{ran}\,\beta$. Then, as in the first part of the proof of 6(d), β is an $\leqq_L$ lower bound of $\{\delta: \alpha\beta \leqq_L \alpha\delta\}$. Now assume that $\alpha^{-1}(\alpha^*(\text{ran}\,\beta)) \neq \text{ran}\,\beta$. Then there are $i \in -\text{ran}\,\beta$ and $k \in \text{ran}\,\beta$ such that $\alpha(i) = \alpha(k)$. By S_t3, $(k/i) \circ \beta$ is in $\mathcal{S}$. Call it δ. Then $\text{ran}(\alpha\beta) = \text{ran}(\alpha\delta)$. By 6(b), $\alpha\beta \leqq_L \alpha\delta$. Since $i \in \text{ran}\,\delta$ and $i \in -\text{ran}\,\beta$, therefore $\text{ran}\,\delta \nsubseteq \text{ran}\,\beta$ and hence $\beta \nleqq_L \delta$. Hence β is not an $\leqq_L$ lower bound of $\{\delta: \alpha\beta \leqq_L \alpha\delta\}$. ♦

Together with Lemma 6, the following lemma and its proof show that many semigroups satisfy S0, ⋯, S9 and all E-conditions other than E_L4.

LEMMA 7: *Each of the semigroups of Theorem* 3 *satisfies* S0, ⋯, S9, E_L2, *and* E_L5.

Proof: Let $\mathcal{S}$ be any of the semigroups of Theorem 3. By induction one proves that if $\mathcal{S}$ is $\text{cl}\{\alpha \in \mathcal{S}_\omega: \alpha\alpha = \alpha\}$ or $\text{cl}\{(0/1), (+1), (-1), (0, 1)\}$ then it is $\{\text{id}\} + \{\alpha \in \mathcal{S}_\omega: \text{ran}\,\alpha \neq I\}$ or $\{\beta\gamma: \beta \in \text{cl}\{(+1), (-1)\} \text{ and } \gamma \in \mathcal{S}_\omega\}$ respectively. One can now verify that $\mathcal{S}$ satisfies S1, $\mathrm{S_t2}$, $\mathrm{S_t3}$, $\mathrm{S_t4}$. By Lemma 6, $\mathcal{S}$ satisfies $\mathrm{E_R1}$, $\mathrm{E_L1}$, $\mathrm{E_R2}$, $\mathrm{E_R3}$. By $\mathrm{S_t2}$ and $\mathrm{E_R2}$, $\mathcal{S}$ satisfies S2. By the proof of Theorem 3, $\mathcal{S}$ satisfies S0.

Now assume that α is idempotent, β is idempotent, $\alpha \leqq_R \beta$, and $\alpha \leqq_L \beta$. Then $\alpha\breve{}\alpha \subseteq \beta\breve{}\beta$ and ran $\beta \subseteq$ ran α. One can verify that the transformation on I which agrees with β on ran α and with the identity transformation on $-$ran α is in $\mathcal{S}$. Call it γ. Since β is idempotent, so is γ. Now consider any j. Then $\alpha(\alpha(j) = \alpha(j)$, hence $\langle\alpha(j), j\rangle \in \alpha\breve{}\alpha \subseteq \beta\breve{}\beta$, and hence $\beta(\alpha(j)) = \beta(j)$. Therefore $\gamma(\alpha(j)) = \beta(\alpha(j)) = \beta(j)$. Thus, $\gamma\alpha = \beta$. Now let i and j be distinct. If both are in $-$ran α, then $\gamma(i) = i$ and $\gamma(j) = j$ are distinct. If i is in $-$ran α and j is in ran α, then $\gamma(i)$ is in $-$ran α and $\gamma(j)$ is in ran $\beta \subseteq$ ran α, so that i and j are again distinct. It follows that $(\gamma\breve{}\gamma) \upharpoonright (\text{ran } \alpha) = \gamma\breve{}\gamma$. By $\mathrm{E_R3}$, γ is an $\leqq_R$ lower bound of $\{\delta: \gamma\alpha \leqq_R \delta\alpha\}$. Hence $\mathcal{S}$ satisfies S5.

Assume that β is idempotent, $\alpha \leqq_L \beta\alpha$, and id is an $\leqq_L$ glb of $\{\alpha, \beta\}$. Then ran $(\beta\alpha) \subseteq$ ran α and, as one can verify, ran α + ran $\beta = I$. Consider any $i \in -$ran α. Then $i \in$ ran β. Hence $\beta(i) = i$. Now consider any $j \neq i$. If $j \in -$ran $\alpha \subseteq$ ran β, then $\beta(j) = j \neq i = \beta(i)$. And if $j \in$ ran α, then $\beta(j) \in$ ran$(\beta\alpha) \subseteq$ ran α so that again $\beta(j) \neq \beta(i) \in -$ran α. It follows that $(\beta\breve{}\beta) \upharpoonright (\text{ran } \alpha) = \beta\breve{}\beta$. By E3, β is an $\leqq_R$ lower bound of $\{\delta: \beta\alpha \leqq_R \delta\alpha\}$. Hence $\mathcal{S}$ satisfies S6.

Consider any α and β. One can verify that $\mathcal{S}$ contains elements α' and β' such that ran $\alpha \subseteq$ ran α', ran $\beta \subseteq$ ran β', ran $\alpha \cdot$ ran β = ran $\alpha' \cdot$ ran β', and ran α' + ran $\beta' = I$. By $\mathrm{E_L1}$ it follows that $\mathcal{S}$ satisfies S7.

Now assume that $\{\alpha, \beta\}$ has some $\leqq_L$ upper bound, i.e., that $\alpha \leqq_L \delta$ and $\beta \leqq_L \delta$ for some δ. Then ran $\delta \subseteq$ ran α and ran $\delta \subseteq$ ran β. Hence ran $\alpha \cdot$ ran $\beta \neq \varnothing$. By S1, $5(\mathrm{c})_\mathrm{L}$, and 6(b), there are α'' and β'' such that α'' and β'' are idempotent, ran α'' = ran α, and ran β'' = ran β. One can verify that there are α' and β' such that α' and α'' agree on ran α'' = ran α, β' and β'' agree on ran β'' = ran β, $\alpha'^*(\text{ran } \beta \cdot -\text{ran } \alpha) \subseteq \text{ran } \alpha \cdot \text{ran } \beta$, $\beta'^*(\text{ran } \alpha \cdot -\text{ran } \beta) \subseteq \text{ran } \alpha \cdot \text{ran } \beta$, and whenever $i \in -\text{ran } \alpha \cdot -\text{ran } \beta$ then $\alpha'(i) = \beta'(i) \in \text{ran } \alpha \cdot \text{ran } \beta$. It follows that $\alpha'\beta' = \beta'\alpha'$, that α' and β' are idempotent, that ran α' =

ran α and ran $\beta' =$ ran β, and hence, by S1, that $\alpha' =_L \alpha$ and $\beta' =_L \beta$. Hence $\mathcal{S}$ satisfies S8.

Assume that γ is an $\leqq_L$ lub of $\{\alpha, \beta\}$. Then $\{\alpha, \beta\}$ has some $\leqq_L$ upper bound. Evidently, ran $\alpha \cdot$ ran $\beta =$ ran$(\alpha'\beta')$ where α' and β' are the transformations which we just constructed for this case. As in the first part of the proof of 6(c) one shows that $\alpha'\beta'$ is an $\leqq_L$ lub of $\{\alpha, \beta\}$. Hence $\gamma \leqq_L \alpha'\beta'$ and $\alpha'\beta' \leqq_L \gamma$. By E_L1, ran $\gamma =$ ran$(\alpha'\beta') =$ ran $\alpha \cdot$ ran β. Conversely if ran $\gamma =$ ran $\alpha \cdot$ ran β then, again as in the first part of the proof of 6(c), γ is an $\leqq$ lub of $\{\alpha, \beta\}$. Hence $\mathcal{S}$ satisfies E_L2.

Clearly, if $\{\alpha, \beta\}$ has an $\leqq_L$ upper bound, then ran $\alpha \cdot$ ran $\beta \neq \varnothing$. Now assume that ran $\alpha \cdot$ ran $\beta \neq \varnothing$. Then, for the α' and β' constructed above, ran$(\alpha'\beta') \subseteq$ ran α and ran$(\alpha'\beta') =$ ran$(\beta'\alpha') \subseteq$ ran β. Then, by E_L1, $\alpha \leqq_L \alpha'\beta'$ and $\beta \leqq_L \alpha'\beta'$. It follows that $\mathcal{S}$ satisfies E_L5.

Consider any α and β. By S_t2, $\mathcal{S}$ contains some δ such that $\delta^{\smile}\delta = \text{eqv}(\alpha^{\smile}\alpha + \beta^{\smile}\beta)$. One can verify that $\mathcal{S}$ contains some γ such that $\gamma^{\smile}\gamma = \text{id} + \alpha^*(\delta^{\smile}\delta)$. Now let $N^0 = \text{id}$ and $N^{n+1} = NN^n$. Since $(\alpha^*\beta^{\smile}\beta))^{\smile} = (\alpha\beta^{\smile}\beta\alpha^{\smile})^{\smile} = \alpha\beta^{\smile}\beta\alpha^{\smile}$, therefore eqv $\alpha^*(\beta^{\smile}\beta)$ is the union of the set $\{(\alpha\beta^{\smile}\beta\alpha^{\smile})^n\colon 0 \leqq n < \omega\}$. Similarly, eqv$(\alpha^{\smile}\alpha + \beta^{\smile}\beta)$ is the union of the set $\{(\alpha^{\smile}\alpha\beta^{\smile}\beta)^n\colon 0 \leqq n < \omega\}$. By induction one can show that

$$\text{id} + \alpha^*(\text{eqv}(\alpha^{\smile}\alpha + \beta^{\smile}\beta)) = \text{eqv}(\alpha^*(\beta^{\smile}\beta)).$$

Hence $\gamma^{\smile}\gamma = \text{eqv}(\alpha^*(\beta^{\smile}\beta))$. Clearly, $(\gamma^{\smile}\gamma) \upharpoonright (\text{ran}\ \alpha) = \gamma^{\smile}\gamma$. Hence, by E_R3, γ is an $\leqq_R$ lower bound of $\{\delta\colon \gamma\alpha \leqq_R \delta\alpha\}$. By induction one can show that $(\gamma\alpha)^{\smile}(\gamma\alpha) = \text{eqv}(\alpha^{\smile}\alpha + \beta^{\smile}\beta)$ and hence that, by E_R2, $\gamma\alpha$ is an $\leqq_R$ lub of $\{\alpha, \beta\}$. Hence $\mathcal{S}$ satisfies S9. ♦

We now return to ^. First we give four conditions on ^ which are applicable whether or not $\mathcal{S}$ is a semigroup of transformations. Obviously, they are equivalent to T_t3, T_t4, T_t5, T_t6 respectively, whenever $\mathcal{S}$ is a semigroup of transformations which satisfies respectively E_R2, E_R3, E_R3, E_R1.

T3. If γ is an $\leqq_R$ lub of $\{\alpha, \beta\}$, then ran $\hat{\alpha} \cdot$ ran $\hat{\beta} =$ ran $\hat{\gamma}$.

T4. If β is an $\leqq_R$ lower bound of $\{\delta\colon \beta\alpha \leqq_R \delta\alpha\}$, then $\hat{\alpha}^{-1}(\hat{\alpha}^*(\text{ran}\,\hat{\beta})) = \text{ran}\,\hat{\beta}$.

T5. If β is not an $\leqq_R$ lower bound of $\{\delta\colon \beta\alpha \leqq_R \delta\alpha\}$, then $-\hat{\alpha}^{-1}(\hat{\alpha}^*(-\text{ran}\,\hat{\beta})) = \text{isl}\mathfrak{J}$.

T6. If ^ is nontrivial, then $\alpha \leqq_R \beta$ if and only if $\text{ran}\,\hat{\beta} \subseteq \text{ran}\,\hat{\alpha}$.

The following theorem follows immediately from Theorem 4 and Lemma 6. It shows that under fairly nonrestrictive conditions on $\mathcal{S}$ the situation described in Theorem 4 can also be described more algebraically. By Lemma 7, this holds in particular for each of the semigroups of Theorem 3.

THEOREM 8: *Assume that* $\mathcal{S}$ *is a semigroup of transformations on* I *which satisfies* S1, S_t2, S_t3, S_t4, *and that* ^ *is the restriction of the* ^ *of* §1 *to* $\mathcal{S}$. *Then* ^ *satisfies* T1, T2, T3, $\cdots$, T6.

Henceforth, in contexts such as $\hat{\alpha} \leqq_L \hat{\beta}$, $\leqq_L$ shall be defined with respect to $\mathfrak{J}$. If ^ is an antirepresentation of $\mathcal{S}$ such that $\alpha \leqq_R \beta$ if and only if $\hat{\alpha} \leqq_L \hat{\beta}$, then ^ shall be $\langle \leqq_R, \leqq_L\rangle$-*faithful*. Evidently, by 6(b), if ^ is a nontrivial antirepresentation of $\mathcal{S}$ and if $\mathfrak{J}$ is regular, then ^ is $\langle \leqq_R, \leqq_L\rangle$-faithful if and only if it satisfies T6.

LEMMA 9: *Assume that* ^ *is a nontrivial antirepresentation of* $\mathcal{S}$, *that* ^ *satisfies* T3, T4, T5, *and that* $\mathcal{S}$ *has an identity and satisfies* S1, S2, S5. *Then* ^ *is* $\langle \leqq_R, \leqq_L\rangle$-*faithful.*

Proof: Assume throughout that ^ satisfies T1 and that $\mathcal{S}$ satisfies S1, i.e., that ^ is an antirepresentation of $\mathcal{S}$ and that $\mathcal{S}$ is regular.

First, note that $\mathfrak{J}$ is regular, since for each $\hat{\alpha}$ there is some β such that $\alpha\beta\alpha = \alpha$ and hence $\hat{\alpha}\hat{\beta}\hat{\alpha} = \widehat{\alpha\beta\alpha} = \hat{\alpha}$. Hence, by 6(b), $\mathfrak{J}$ also satisfies E_L1.

Assume for a moment that $\alpha \leqq_R \beta$. Then $\gamma\alpha = \beta$ for some γ. Then $\hat{\alpha}\hat{\gamma} = \widehat{\gamma\alpha} = \hat{\beta}$. Hence $\hat{\alpha} \leqq_L \hat{\beta}$. Thus, if $\alpha \leqq_R \beta$ then $\hat{\alpha} \leqq_L \hat{\beta}$.

Assume from now on that ^ is nontrivial and satisfies T5 and that $\mathcal{S}$ has an identity id. Consider any idempotent δ distinct from id. Then $\delta \not\leqq_R \text{id}$ since, by $5(\text{a})_R$, $\delta \leqq_R \text{id}$ would imply that $\delta = \text{id} \circ \delta = \text{id}$. Also $\text{id} \circ \delta\delta = \text{id} \circ \delta$ and hence $\delta\delta \leqq_R \text{id} \circ \delta$. Therefore δ is not an $\leqq_R$ lower bound of $\{\gamma\colon \delta\delta \leqq_R \gamma\delta\}$. By T5, $-\hat{\delta}^{-1}(\hat{\delta}^*(-\text{ran}\,\hat{\delta})) = \text{isl}\mathfrak{J}$. Since ^ and hence $\mathfrak{J}$ is nontrivial, dom $\hat{\delta} \nsubseteq$ isl$\mathfrak{I}$. It follows that $\hat{\delta}$ is not the identity transformation on dom $\hat{\delta}$. Since $\hat{\delta}\hat{\delta} = \widehat{\delta\delta} = \hat{\delta}$, it follows that dom $\hat{\delta} \nsubseteq$ ran $\hat{\delta}$.

Assume now also that ^ satisfies T4 and that $\mathcal{S}$ satisfies S5. Consider any β and γ such that $\beta \leqq_R \gamma$ and $\gamma \nleqq_R \beta$. Consider first the case where β and γ are idempotent and where $\beta \leqq_L \gamma$. By S5, there is some δ which is idempotent such that $\widehat{\delta\beta} = \gamma$ and δ is an $\leqq_R$ lower bound of $\{\eta: \delta\beta \leqq_R \eta\beta\}$. Then $\delta \nleqq_R$ id since otherwise it would follow, by $5(\text{a})_{\text{R}}$, that id $\circ\, \delta =$ id, hence that id $\circ\, \gamma =$ id $\circ\, \delta\beta =$ id $\circ\, \beta = \beta$, and hence that $\gamma \leqq_R \beta$. Also, by T4 and T1, ran $\hat{\delta} = \hat{\beta}^{-1}(\hat{\beta}^*(\text{ran } \hat{\delta})) = \hat{\beta}^{-1}(\text{ran}(\hat{\beta}\hat{\delta})) = \hat{\beta}^{-1}(\text{ran } \widehat{\delta\beta}) = \hat{\beta}^{-1}(\text{ran } \hat{\gamma})$. Then ran $\hat{\gamma} \neq$ ran $\hat{\beta}$ since otherwise it would follow that ran $\hat{\delta} = \hat{\beta}^{-1}(\text{ran } \hat{\gamma}) = \hat{\beta}^{-1}(\text{ran } \hat{\beta}) =$ dom $\hat{\beta} =$ dom $\hat{\delta}$. For the remaining case it follows from the previous case and from the regularity of $\mathfrak{I}$ and $5(\text{d})_{\text{R}}$ that again ran $\hat{\gamma} \neq$ ran $\hat{\beta}$.

Assume now also that ^ satisfies T3 and that $\mathcal{S}$ satisfies S2. Consider any α and β such that $\alpha \nleqq_R \beta$. By S2, there is some γ such that γ is an $\leqq_R$ lub of $\{\alpha, \beta\}$. Then $\beta \leqq_R \gamma$ and $\gamma \nleqq_R \beta$. By T3, ran $\hat{\gamma} =$ ran $\hat{\alpha} \cdot$ ran $\hat{\beta}$. Since ran $\hat{\beta} \subseteq$ ran $\hat{\alpha}$ would imply that ran $\hat{\gamma} =$ ran $\hat{\beta}$, it follows that ran $\hat{\beta} \nsubseteq$ ran $\hat{\alpha}$ and hence that $\hat{\alpha} \nleqq_L \hat{\beta}$. Therefore, if $\hat{\alpha} \leqq_L \hat{\beta}$ then $\alpha \leqq_R \beta$. ♦

We now show that T3, T4, T5 can often be replaced by conditions of which only one involves $\mathcal{S}$, while the others are internal requirements on $\mathfrak{I}$.

THEOREM 10: *Assume that ^ is a nontrivial antirepresentation of* $\mathcal{S}$ *and that* $\mathcal{S}$ *has an identity and satisfies* S1, S2, S5. *Then ^ satisfies* T3, T4, *and* T5 *if and only if ^ is* $\langle \leqq_R, \leqq_L \rangle$*-faithful and* $\mathfrak{I}$ *has suitable ranges.*

Proof: Assume the hypotheses of the theorem and assume that ^ is $\langle \leqq_R, \leqq_L \rangle$-faithful. By Lemma 9, it suffices to show that ^ satisfies T3, T4, and T5 if and only if $\mathfrak{I}$ satisfies $\text{E}_\text{L}2$, $\text{E}_\text{L}3$, and $\text{E}_\text{L}4$.

From the $\langle \leqq_R, \leqq_L \rangle$-faithfulness of ^ there follow these two equivalences: (i) γ is an $\leqq_R$ lub of $\{\alpha, \beta\}$ if and only if $\hat{\gamma}$ is an $\leqq_L$ lub of $\{\hat{\alpha}, \hat{\beta}\}$; (ii) β is an $\leqq_R$ lower bound of $\{\delta: \beta\alpha \leqq_R \delta\alpha\}$ if and only if $\hat{\beta}$ is an $\leqq_L$ lower bound of $\{\hat{\delta}: \hat{\alpha}\hat{\beta} \leqq_L \hat{\alpha}\hat{\delta}\}$.

Assume for a moment that $\mathfrak{J}$ satisfies E_L2. Suppose that γ is an $\leqq_R$ lub of $\{\alpha, \beta\}$. By (i), $\hat{\gamma}$ is an $\leqq_L$ lub of $\{\hat{\alpha}, \hat{\beta}\}$. By E_L2, $\operatorname{ran}\hat{\alpha} \cdot \operatorname{ran}\hat{\beta} = \operatorname{ran}\hat{\gamma}$. Hence $\hat{}$ satisfies T3. Hence, if $\mathfrak{J}$ satisfies E_L2, then $\hat{}$ satisfies T3. Similarly, by (ii), if $\mathfrak{J}$ satisfies E_L3 then $\hat{}$ satisfies T4, and if $\mathfrak{J}$ satisfies E_L4, then $\hat{}$ satisfies T5.

For the rest of the proof, assume that $\hat{}$ satisfies T3, T4, and T5. By T4 and T5, either $\hat{\alpha}^{-1}(\hat{\alpha}^*(\operatorname{ran}\hat{\beta})) = \operatorname{ran}\hat{\beta}$ or $-\hat{\alpha}^{-1}(\hat{\alpha}^*(-\operatorname{ran}\hat{\beta})) = \operatorname{isl}\mathfrak{J}$. Also, either not $\hat{\alpha}^{-1}(\hat{\alpha}^*(\operatorname{ran}\hat{\beta})) = \operatorname{ran}\hat{\beta}$ or not $-\hat{\alpha}^{-1}(\hat{\alpha}^*(-\operatorname{ran}\hat{\beta})) = \operatorname{isl}\mathfrak{J}$ since, if both were true then $\operatorname{ran}\hat{\beta} = -\hat{\alpha}^{-1}(\hat{\alpha}^*(-\operatorname{ran}\hat{\beta})) = \operatorname{isl}\mathfrak{J}$, which implies that $\hat{\beta}$ is the identity transformation on $\operatorname{dom}\hat{\beta}$, hence that $\operatorname{dom}\hat{\beta} = \operatorname{ran}\hat{\beta} = \operatorname{isl}\mathfrak{J}$, and hence that $\mathfrak{J}$ is trivial. Thus, $\hat{\alpha}^{-1}(\hat{\alpha}^*(\operatorname{ran}\hat{\beta})) = \operatorname{ran}\hat{\beta}$ and $-\hat{\alpha}^{-1}(\hat{\alpha}^*(-\operatorname{ran}\hat{\beta})) \neq \operatorname{isl}\mathfrak{J}$ are equivalent. From T4, T5, and (ii) there follow E_L3 and E_L4.

From T3 and (i) it follows that if $\hat{\gamma}$ is an $\leqq_L$ lub of $\{\hat{\alpha}, \hat{\beta}\}$ then $\operatorname{ran}\hat{\alpha} \cdot \operatorname{ran}\hat{\beta} = \operatorname{ran}\hat{\gamma}$. Now suppose that $\operatorname{ran}\hat{\alpha} \cdot \operatorname{ran}\hat{\beta} = \operatorname{ran}\hat{\gamma}$. As we saw in the proof of Lemma 9, $\mathfrak{J}$ is regular and hence satisfies E_L1. It follows that $\hat{\alpha} \leqq_L \hat{\gamma}$ and $\hat{\beta} \leqq_L \hat{\gamma}$. Also, if $\hat{\alpha} \leqq_L \hat{\delta}$ and $\hat{\beta} \leqq_L \hat{\delta}$, then $\operatorname{ran}\hat{\delta} \subseteq \operatorname{ran}\hat{\alpha}$, $\operatorname{ran}\hat{\delta} \subseteq \operatorname{ran}\hat{\beta}$, hence $\operatorname{ran}\hat{\delta} \subseteq \operatorname{ran}\hat{\alpha} \cdot \operatorname{ran}\hat{\beta} = \operatorname{ran}\hat{\gamma}$, and therefore, by E_L1, $\hat{\gamma} \leqq_L \hat{\delta}$. Hence $\hat{\gamma}$ is an $\leqq_L$ lub of $\{\hat{\alpha}, \hat{\beta}\}$. Therefore $\mathfrak{J}$ satisfies E_L2. ♦

3. TRANS-BOOLEAN ALGEBRAS. ELEMENTARY PROPERTIES.

We shall begin §3 by abstracting from the transformational operations on sets of sequences considered in §1 and from the Boolean operations on such sets. In the remainder of the paper we shall investigate the resulting algebraic theory. In the rest of §3, we shall derive some elementary consequences, indicating in each case which axioms, or assumptions, are used. For many of these consequences the assumptions used are fairly general. Also, they involve only algebraic conditions concerning $\mathcal{S}$, and do not require that $\mathcal{S}$ is a semigroup of transformations.

We continue to let $\mathcal{S} = \langle S, \circ \rangle$ be a semigroup which remains fixed. We also continue to let $\alpha, \beta, \gamma, \cdots$ be arbitrary elements of

$\mathcal{S}$ and id the identity element of $\mathcal{S}$, if $\mathcal{S}$ has such an element. Also, $\mathcal{S}$ and $\mathfrak{I}$ shall continue to be the domain and range respectively of $\hat{}$. In contrast to §2, however, $\hat{}$ may vary.

An *algebra* shall be any $\mathfrak{A} = \langle A, +, \cdot, -, 1, (s\alpha), (t\alpha) \rangle_{\alpha \in \mathcal{S}}$ such that A is a nonempty set and such that $+, \cdot, -, 1$, and each $(s\alpha)$ and $(t\alpha)$ are operations on A taking 2, 2, 1, 0, 1, 1 arguments respectively. Later on, in §4, we let other structures be algebras also. In particular, when $\mathcal{S}$ is a semigroup of transformations, we shall at times consider algebras

$$\langle A, +, \cdot, -, 1, (s\alpha), (t\alpha), (cJ), (dM) \rangle_{\alpha \in \mathcal{S}, J \in \mathcal{C}(\mathcal{S}), M \in \mathcal{E}(\mathcal{S})}$$

where each (cJ) and (dM) takes respectively 1 or 0 arguments. Given an algebra $\mathfrak{A}$, we let $\varnothing$ be the element -1, and we let $\leqq$ be the relation $\{\langle x, y \rangle \in A \times A : x + y = y\}$ and $\geqq$ its converse. Henceforth $x, y, z, \cdots$ shall be arbitrary elements of the set A of the algebra $\mathfrak{A}$ under consideration.

Let $\hat{}$ be any prerepresentation of $\mathcal{S}$, and let $W = \text{dom}\, \hat{\alpha}$ for some α (and hence for each α). The *complete algebra induced by* $\hat{}$ shall be that algebra $\mathfrak{A} = \langle A, +, \cdot, -, 1, (s\alpha), (t\alpha) \rangle_{\alpha \in \mathcal{S}}$ such that A is $\{x : x \subseteq W\}$, $+$ is union, $\cdot$ is intersection, $-$ is complementation with respect to W, 1 is W, $(s\alpha)x$ is $\hat{\alpha}^{-1}x = \{w \in W : \hat{\alpha}(w) \in x\}$, and $(t\alpha)x$ is $\hat{\alpha}^{*}x = \{\hat{\alpha}(w) : w \in x\}$ for each $x \in A$. Any subalgebra of $\mathfrak{A}$ shall be *induced by* $\hat{}$. If $\mathcal{S}$ is a semigroup of transformations on I, $\hat{}$ is the restriction of the $\hat{}$ of §1, and $\mathfrak{A}$ is the complete algebra induced by $\hat{}$, then $+, \cdot, -, 1, (s\alpha), (t\alpha)$ are those operations on sets of sequences which we considered in §1.

Below we shall give sets TB0, $\cdots$, TB10 of conditions on an algebra $\mathfrak{A} = \langle A, +, \cdot, -, 1, (s\alpha), (t\alpha) \rangle_{\alpha \in \mathcal{S}}$. By our definition of $\leqq$, each condition in one of these sets may be regarded as an equality. These equalities will be used as axioms for an abstract theory of transformational operations (resulting from $\mathcal{S}$) and of Boolean operations. Any $\mathfrak{A}$ satisfying all of these equalities shall be a *trans-Boolean* algebra. We let $\text{TB}_{\mathcal{S}}$ be the set of all of these equalities, and $\text{TB}_{\mathcal{S}}^{-}$ the set of those which belong to one of the sets TB0, $\cdots$, TB9. We shall say, for example, that $\mathfrak{A}$ *satisfies* TB8 if and only if, for any elements of A, each equality in TB8 is true for $\mathfrak{A}$.

Most of the sets TB0, $\cdots$, TB10 now to be given have been taken or adapted from the literature. Specifically, TB1, TB2, and TB6 have been taken from p. 111 of [4], TB4 and TB7 correspond to (C_2) and (C_4) respectively of §1.1 of [5], TB8 is adapted from (D4) on p. 260 of [9], and TB9 is adapted from a special case of (C3) on p. 240 of [9]. The requirement in TB9 on α and β (as well as the negation of the requirement in TB10) can also be expressed as follows: For every δ, if $\beta\alpha \leqq_R \delta\alpha$, then $\beta \leqq_R \delta$.

TB0. $\langle A, +, \cdot, -, \mathit{1} \rangle$ is a Boolean algebra.
TB1. $(s\alpha) - x = -(s\alpha)x$.
TB2. $(s\alpha)(x + y) = (s\alpha)x + (s\alpha)y$.
TB3. $(t\alpha)(x + y) = (t\alpha)x + (t\alpha)y$.
TB4. $(s\alpha)(t\alpha)x \geqq x$.
TB5. $(t\alpha)(s\alpha)x \leqq x$.
TB6. $(s\alpha)(s\beta)x = (s\alpha\beta)x$.
TB7. $(s\alpha)(t\alpha)(s\beta)(t\beta)x = (s\beta)(t\beta)(s\alpha)(t\alpha)x$.
TB8. $(t\alpha)\mathit{1} \cdot (t\beta)\mathit{1} \leqq (t\gamma)\mathit{1}$, provided γ is an $\leqq_R$ lub of $\{\alpha, \beta\}$.
TB9. $(s\alpha)(t\alpha)(t\beta)\mathit{1} = (t\beta)\mathit{1}$, provided β is an $\leqq_R$ lower bound of $\{\delta : \beta\alpha \leqq_R \delta\alpha\}$.
TB10. $(s\gamma)(x \cdot -(s\alpha)(t\alpha) - (t\beta)\mathit{1}) = x \cdot -(s\alpha)(t\alpha) - (t\beta)\mathit{1}$, provided β is not an $\leqq_R$ lower bound of $\{\delta : \beta\alpha \leqq_R \delta\alpha\}$.

THEOREM 11: *Assume that* $\mathfrak{A}$ *is induced by a prerepresentation* $\hat{}$ *of* $\mathcal{S}$.

(a) $\mathfrak{A}$ *satisfies* TB0, $\cdots$, TB5.
(b) *If* $\hat{}$ *satisfies* T1, *then* $\mathfrak{A}$ *satisfies* TB6.
(c) *If* $\hat{}$ *satisfies* T2, *then* $\mathfrak{A}$ *satisfies* TB7.
(d) *If* $\hat{}$ *satisfies* T3, *then* $\mathfrak{A}$ *satisfies* TB8.
(e) *If* $\hat{}$ *satisfies* T4, *then* $\mathfrak{A}$ *satisfies* TB9.
(f) *If* $\hat{}$ *satisfies* T5, *then* $\mathfrak{A}$ *satisfies* TB10.

Proof: (f) Assume that $\hat{}$ is a prerepresentation of $\mathcal{S}$ which satisfies T5, and that β is not an $\leqq_R$ lower bound of $\{\delta : \beta\alpha \leqq_R \delta\alpha\}$. Let $\mathfrak{J}$ be the range of $\hat{}$, and let $y = x \cdot -(s\alpha)(t\alpha) - (t\beta)\mathit{1} = x \cdot -\hat{\alpha}^{-1}(\hat{\alpha}^*(-\operatorname{ran} \hat{\beta}))$. Then $y \subseteq \operatorname{isl}\mathfrak{J}$. Hence $\hat{\gamma}^{-1}\{w\} = \{w\}$ for each $w \in y$. Hence $(s\gamma)y = \hat{\gamma}^{-1}y = y$. ♦

The next theorem, which follows from Theorem 8 and Theorem

11, shows that, for many semigroups $\mathcal{S}$ of transformations, our axioms are satisfied by the set-theoretic transformational and Boolean operations.

THEOREM 12: *Assume that $\mathcal{S}$ is a semigroup of transformations on I which satisfies* S1, S_t2, S_t3, S_t4, *that* ^ *is the restriction of the* ^ *of* §1 *to* $\mathcal{S}$, *and that* $\mathfrak{A}$ *is induced by* ^. *Then* $\mathfrak{A}$ *satisfies* TB0, ⋯, TB10.

In the rest of §3, we shall derive elementary consequences of certain subsets of TB0, ⋯, TB9. These consequences hold, in particular, for any $\mathfrak{A}$ which satisfies the appropriate hypotheses in Theorem 11. First, we consider consequences of TB0, ⋯, TB5.

LEMMA 13: *Assume that* $\mathfrak{A}$ *satisfies* TB0, ⋯, TB5.

(a) *If* $x \leqq y$, *then* $(s\alpha)x \leqq (s\alpha)y$.
(b) *If* $x \leqq y$, *then* $(t\alpha)x \leqq (t\alpha)y$.
(c) $(s\alpha)\varnothing = \varnothing$.
(d) $(s\alpha)\mathcal{I} = \mathcal{I}$.
(e) $(s\alpha)(x \cdot y) = (s\alpha)x \cdot (s\alpha)y$.
(f) $(t\alpha)\varnothing = \varnothing$.
(g) $(s\alpha)(t\alpha)(s\alpha)x = (s\alpha)x$.
(h) $(t\alpha)(s\alpha)(t\alpha)x = (t\alpha)x$.
(i) $(s\alpha)(t\alpha)\varnothing = \varnothing$.

Proof: Henceforth, TB0 will often be used tacitly.

(a) By TB2.
(b) By TB3.
(c), (d), (e) By TB1 and TB2.
(f) By 13(c) and TB5 respectively one has $(t\alpha)\varnothing = (t\alpha)(s\alpha)\varnothing \leqq \varnothing$.
(g) By TB4 one has $(s\alpha)(t\alpha)(s\alpha)x \geqq (s\alpha)x$. By TB5 one has $x \leqq (t\alpha)(s\alpha)x$, which, with 13(a), yields $(s\alpha)x \geqq (s\alpha)(t\alpha)(s\alpha)x$.
(h) By TB5 one has $(t\alpha)(s\alpha)(t\alpha)x \leqq (t\alpha)x$. By TB4 one has $x \leqq (s\alpha)(t\alpha)x$, which, with 13(b), yields $(t\alpha)x \leqq (t\alpha)(s\alpha)(t\alpha)x$.
(i) By 13(f) and 13(c). ♦

Part (e) of the following lemma corresponds to (C_3) of [5], while 13(i) and TB4 correspond to (C_1) and (C_2) respectively. Part (f) mirrors the earlier definition of $(e\alpha\breve{}\alpha)x$ as $x \cdot (d\alpha\breve{}\alpha)$.

LEMMA 14: *Assume that* $\mathfrak{A}$ *satisfies* TB0, $\cdots$, TB5.

(a) *If* $x \leqq (s\alpha)y$, *then* $(t\alpha)x \leqq y$.
(b) *If* $(t\alpha)x \leqq y$, *then* $x \leqq (s\alpha)y$.
(c) $(t\alpha)x \cdot y = \emptyset$ *if and only if* $x \cdot (s\alpha)y = \emptyset$.
(d) $(t\alpha)(x \cdot (s\alpha)y) = (t\alpha)x \cdot y$.
(e) $(s\alpha)(t\alpha)(x \cdot (s\alpha)(t\alpha)y) = (s\alpha)(t\alpha)x \cdot (s\alpha)(t\alpha)y$.
(f) $(t\alpha)(s\alpha)y = (t\alpha)\mathit{1} \cdot y$.
(g) $(t\alpha)(s\alpha)(t\beta)(s\beta)x = (t\beta)(s\beta)(t\alpha)(s\alpha)x$.
(h) $(s\alpha)(t\alpha)x = x$ *if and only if* $-(s\alpha)(t\alpha) - x = x$.

Proof:

(a) Assume that $x \leqq (s\alpha)y$. Then $(t\alpha)x \leqq (t\alpha)(s\alpha)y \leqq y$ by 13(b) and TB5 respectively.

(b) Assume that $y \geqq (t\alpha)x$. Then $(s\alpha)y \geqq (s\alpha)(t\alpha)x \geqq x$ by 13(a) and TB4 respectively.

(c) $(t\alpha)x \cdot y = \emptyset$ iff $(t\alpha)x \leqq -y$ iff $x \leqq (s\alpha) - y$ iff $x \leqq -(s\alpha)y$ iff $x \cdot (s\alpha)y = \emptyset$ by TB0, 14(a) and 14(b), TB1, and TB0 respectively.

(d) Since $x \cdot (s\alpha)y \leqq (s\alpha)y$, therefore $(t\alpha)(x \cdot (s\alpha)y) \leqq (t\alpha)(s\alpha)y \leqq y$ by 13(b) and TB5 respectively. Likewise, $(t\alpha)(x \cdot (s\alpha)y) \leqq (t\alpha)x$. Hence $(t\alpha)(x \cdot (s\alpha)y) \leqq (t\alpha)x \cdot y$.

By TB4 one has $x \cdot (s\alpha)y \leqq (s\alpha)(t\alpha)(x \cdot (s\alpha)y)$. Hence $x \cdot (s\alpha)(y \cdot -(t\alpha)(x \cdot (s\alpha)y)) = \emptyset$, by TB0, TB1, and TB2. Then $(t\alpha)x \cdot (y \cdot -(l\alpha)(x \cdot (s\alpha)y)) = \emptyset$ by 14(c), and hence $(t\alpha)x \cdot y \leqq (t\alpha)(x \cdot (s\alpha)y)$.

(e) $(s\alpha)(t\alpha)(x \cdot (s\alpha)(t\alpha)y) = (s\alpha)((t\alpha)x \cdot (t\alpha)y) = (s\alpha)(t\alpha)x \cdot (s\alpha)(t\alpha)y$ by 14(d) and 13(e) respectively.

(f) $(t\alpha)(s\alpha)y = (t\alpha)(\mathit{1} \cdot (s\alpha)y) = (t\alpha)\mathit{1} \cdot y$ by TB0 and 14(d) respectively.

(g) By 14(f) and TB0.

(h) Using TB0 and respectively TB4 and 14(c) one sees that $(s\alpha)(t\alpha)x = x$, $(s\alpha)(t\alpha)x \leqq x$, and $(t\alpha)x \cdot (t\alpha) - x = 0$

are equivalent. Using TB0 and respectively TB4 and 14(c) one also sees that $-(s\alpha)(t\alpha) - x = x$, $-(s\alpha)(t\alpha) - x \geqq x$, and $(t\alpha)x \cdot (t\alpha) - x = 0$ are equivalent. ♦

Given TB0, TB4 and 14(a) together have a simple meaning. Let $\mathfrak{A} = \langle A, +, \cdot, -, 1, (s\alpha), (t\alpha) \rangle_{\alpha \in \mathcal{S}}$ be such that $\langle A, +, \cdot, -, 1 \rangle$ is a Boolean algebra. Then $\mathfrak{A}$ satisfies TB4 and 14(a) if and only if, for any $x \in A$, $(t\alpha)x$ is the least element of $\{y\colon (s\alpha)y \geqq x\}$. Likewise, $\mathfrak{A}$ satisfies TB5 and 14(b) if and only if, for any $x \in A$, $(s\alpha)x$ is the greatest element of $\{y\colon (t\alpha)y \leqq x\}$.

Conditions TB3 and TB5 may be replaced by 14(a). First, from 14(a) one can derive that $(s\alpha)x \leqq (s\alpha)x$ implies $(t\alpha)(s\alpha)x \leqq x$. Hence 14(a) and TB0 yield TB5. Also, apart from uses of TB0, the proof of 14(c) used only 14(a), TB1, and 14(b), that of 14(b) only TB4 and 13(a), and that of 13(a) only TB2. Hence TB0, TB1, TB2, TB4, and 14(a) yield 14(c). But 14(c) and TB0 yield TB3, by 15(c) below.

One may regard TB4 and 14(a) together as a definition of $(t\alpha)$ and construct an axiom set essentially equivalent to TB0, TB1, TB2, TB4, 14(a) and hence to TB0, TB1, TB2, TB3, TB4, TB5, whose only extra-Boolean primitives are the operations $(s\alpha)$, $\alpha \in \mathcal{S}$. More precisely, for each $\alpha \in \mathcal{S}$ one adds to TB0, TB1, TB2 the condition: For each $x \in A$ there is a least z such that $(s\alpha)z \geqq x$. Similar remarks apply to TB0, $\cdots$, TB10. This amounts to adapting the interdefinability of conjugates described in [7] to the present situation.

In passing we note an alternative axiom set. We saw that, TB0 and 14(c) yield TB3. They likewise yield TB2. Also, TB0, TB1, and 14(c) yield 14(a) and 14(b). Finally, with TB0, as we saw, 14(a) yields TB5 and similarly 14(b) yields TB4. Hence one may replace TB0, $\cdots$, TB5 by TB0, TB1, 14(c).

Another set which may be used instead is TB0, TB1, TB4, 13(d), 13(f), 14(d), from which 14(c) can be derived. This set, which is closely related to Howard's axiomatization in [6], is made up of equalities.

A further set which may be used is TB0, TB1, 13(a), 13(b), TB4, TB5. (The last four of these are meaningful for any partial ordering.) For proof note that our derivation of 14(c) used only TB0, TB1, 14(a), 14(b), that 14(a) was derived from 13(b) and TB5, and that 14(b) was derived from 13(a) and TB4.

Lemmas 13(g) and 14(g), for example, involve only 1-ary operations. Using (as before) juxtaposition to indicate composition of functions, we shall often render them more concisely as $(s\alpha)(t\alpha)(s\alpha) = (s\alpha)$ and $(t\alpha)(s\alpha)(t\beta)(s\beta) = (t\beta)(s\beta)(t\alpha)(s\alpha)$ respectively. A similar notation for inequalities will be useful. Henceforth we often let, for example, $(s\alpha)(t\alpha) \leqq (s\beta)(t\beta)$ if and only if $(s\alpha)(t\alpha)x \leqq (s\beta)(t\beta)x$ for each $x \in A$. We also let, for example, $(s\beta)(t\beta) \geqq (s\alpha)(t\alpha)$ if and only if $(s\alpha)(t\alpha) \leqq (s\beta)(t\beta)$. We shall freely change from one notation to the other.

Let $\mathfrak{A} = \langle A, +, \cdot, -, 1\rangle$ be a Boolean algebra. Then F and F' shall be *conjugate* (in $\mathfrak{A}$) if and only if F and F' are 1-ary operations on A such that, for any x and y in A, $Fx \cdot y = \emptyset$ if and only if $x \cdot F'y = \emptyset$. By 14(c), if $\langle A, +, \cdot, -, 1, (s\alpha), (t\alpha)\rangle_{\alpha\in S}$ also satisfies TB1, $\cdots$, TB5, then any $(s\alpha)$ and $(t\alpha)$ are conjugate.

Three simple facts concerning conjugates will be useful.

LEMMA 15: *Assume that $\mathfrak{A} = \langle A, +, \cdot, -, 1\rangle$ is a Boolean algebra, that F and F' are conjugate, and that G and G' are conjugate.*

(a) *FG and $G'F'$ are conjugate.*

(b) *If $F \leqq G$, then $F' \leqq G'$. Hence if $F = G$, then $F' = G'$.*

(c) *For any $B \subseteq A$, if B has a least upper bound $z \in A$, then Fz is the least upper bound of $\{Fy: y \in B\}$. In particular, if $B = \{x, y\}$ and $z = x + y$, then $F(x + y) = Fz = Fx + Fy$.*

Proof:

(a) For any x and y in A, $FGx \cdot y = \emptyset$ iff $Gx \cdot F'y = \emptyset$ iff $x \cdot G'F'y = \emptyset$.

(b) Assume that $Fx \leqq Gx$ for each $x \in A$. Consider any $y \in A$. Suppose that $x \cdot G'y = \emptyset$. Then $Gx \cdot y = \emptyset$. By our assumption, $Fx \cdot y = \emptyset$. Hence $x \cdot F'y = \emptyset$. In particular, if $x = -G'y$, then $x \cdot G'y = -G'y \cdot G'y = \emptyset$. Hence $-G'y \cdot Gy = \emptyset$. Since y is arbitrary, it follows that $G \leqq G'$.

(c) We use the argument of [7], p. 904. Assume that $B \subseteq A$

has a least upper bound $z \in A$. Assume that x is an upper bound of $\{Fy: y \in B\}$, i.e., that $Fy \leqq x$ for each $y \in B$. Then $Fy \cdot -x = \emptyset$ and hence, by conjugacy, $y \cdot G - x = \emptyset$ for each $y \in B$. Then $z \cdot G - x = \emptyset$ and hence, by conjugacy, $Fz \cdot -x = \emptyset$. Hence $Fz \leqq x$. Each of these steps is reversible, so that $Fz \leqq x$ iff x is an upper bound of $\{Fy: y \in B\}$. ♦

We now turn to consequences of TB0, ⋯, TB5, TB6. Part (g) of the following lemma corresponds to the axiom scheme of substitutivity on pp. 215–216 of Halmos [4]. The derivation of part (d) from part (c) illustrates a general method.

LEMMA 16: *Assume that* $\mathfrak{A}$ *satisfies* TB0, ⋯, TB6.

(a) $(t\beta)(t\alpha) = (t\alpha\beta)$.
(b) $(t\beta')(s\alpha') \leqq (s\alpha)(t\beta)$, *provided* $\beta'\alpha = \alpha'\beta$.
(c) $(s\beta)(t\beta)(s\alpha) \leqq (s\alpha)(s\gamma)(t\gamma)$, *provided* $\beta\alpha = \alpha\gamma$.
(d) $(t\alpha)(s\beta)(t\beta) \leqq (s\gamma)(t\gamma)(t\alpha)$, *provided* $\beta\alpha = \alpha\gamma$.
(e) $(s\alpha)(t\alpha)((t\alpha)\mathfrak{1} \cdot x) = (s\alpha)x$, *provided* $\alpha\alpha = \alpha$.
(f) $(t\alpha)\mathfrak{1} \cdot (s\alpha)(t\alpha)x = (t\alpha)x$, *provided* $\alpha\alpha = \alpha$.
(g) $(t\alpha)\mathfrak{1} \cdot x \leqq (s\alpha)x$, *provided* $\alpha\alpha = \alpha$.

Proof:

(a) By TB6, 15(a), and 15(b).

(b) Assume that $\beta'\alpha = \alpha'\beta$. Then $(t\beta')(s\alpha')x \leqq (s\alpha)(t\beta)x$ iff $(t\beta')(s\alpha')x \cdot -(s\alpha)(t\beta)x = \emptyset$ iff $(t\beta')(s\alpha')x \cdot (s\alpha) - (t\beta)x = \emptyset$ iff $(t\alpha)(t\beta')(s\alpha')x \cdot -(t\beta)x = \emptyset$ iff $(t\beta'\alpha)(s\alpha')x \cdot -(t\beta)x = \emptyset$ iff $(t\alpha'\beta)(s\alpha')x \cdot -(t\beta)x = \emptyset$ iff $(t\beta)(t\alpha')(s\alpha')x \cdot -(t\beta)x = \emptyset$ iff $(t\beta)(t\alpha')(s\alpha')x \leqq (t\beta)x$ by TB0, TB1, 14(c), 16(a), $\beta'\alpha = \alpha'\beta$, 16(a), and TB0 respectively. Now $(t\beta)(t\alpha')(s\alpha')x \leqq (t\beta)x$, by TB5 and 13(b). Hence $(t\beta')(s\alpha')x \leqq (s\alpha)(t\beta)x$.

(c) Assume that $\beta\alpha = \alpha\gamma$. By 16(b), $(t\beta)(s\alpha) \leqq (s\alpha)(t\gamma)$. Then $(s\beta)(t\beta)(s\alpha) \leqq (s\beta)(s\alpha)(t\gamma) = (s\beta\alpha)(t\gamma) = (s\alpha\gamma)(t\gamma) = (s\alpha)(s\gamma)(t\gamma)$ by 13(a), TB6, $\beta\alpha = \alpha\gamma$, and TB6 respectively.

(d) By 16(c), 14(c), 15(a), and 15(b).

(e) $(s\alpha)x = (s\alpha)(t\alpha)(s\alpha)x = (s\alpha)(t\alpha)(t\alpha)(s\alpha)x = (s\alpha)(t\alpha)((t\alpha)\mathfrak{1} \cdot x)$ by 13(g), $\alpha\alpha = \alpha$ and 16(a), and 14(f) respectively.

(f) $(t\alpha)x = (t\alpha)(s\alpha)(t\alpha)x = (t\alpha)(s\alpha)(s\alpha)(t\alpha)x = (t\alpha)\mathfrak{1} \cdot (s\alpha)(t\alpha)x$ by 13(h), $\alpha\alpha = \alpha$ and TB6, and 14(f) respectively.

(g) By 16(e) and TB4. ♦

Some effects of regularity will now be described. Parts (a), (c), (d) of the following lemma relate $\leqq$ for $(s\alpha)(t\alpha)$, $(t\alpha)(s\alpha)$, and $(t\alpha)\mathfrak{1}$ respectively to divisibility in the semigroup of operations $(s\alpha)$ under composition. Part (b) implies that $\alpha =_L \beta$ is a sufficient condition for $(s\alpha)(t\alpha) = (s\beta)(t\beta)$. Part (e) implies that $\alpha =_R \beta$ is a sufficient condition for $(t\alpha)\mathfrak{1} = (t\beta)\mathfrak{1}$. The if-portions of (a), (c), (d) do not require TB6 or regularity.

LEMMA 17: *Assume that* $\mathfrak{A}$ *satisfies* TB0, $\cdots$, TB6 *and that* $\mathcal{S}$ *is regular.*

(a) $(s\alpha)(t\alpha) \geqq (s\beta)(t\beta)$ *if and only if there is some* γ *such that* $(s\alpha) = (s\beta)(s\gamma)$.

(b) $(s\alpha)(t\alpha) \leqq (s\beta)(t\beta)$, *provided* $\alpha \leqq_L \beta$.

(c) $(t\alpha)(s\alpha) \leqq (t\beta)(s\beta)$ *if and only if there is some* γ *such that* $(s\alpha) = (s\gamma)(s\beta)$.

(d) $(t\alpha)\mathfrak{1} \leqq (t\beta)\mathfrak{1}$ *if and only if there is some* γ *such that* $(s\alpha) = (s\gamma)(s\beta)$.

(e) $(t\alpha)\mathfrak{1} \leqq (t\beta)\mathfrak{1}$, *provided* $\beta \leqq_R \alpha$.

(f) $(s\alpha)(t\beta)\mathfrak{1} \geqq (t\gamma)\mathfrak{1}$, *provided* $\beta \leqq_R \gamma\alpha$.

Proof:

(a) Assume that $(s\alpha) = (s\beta)(s\gamma)$. By 15(a) and 15(b), $(t\alpha) = (t\gamma)(t\beta)$. Hence $(s\alpha)(t\alpha) = (s\beta)(s\gamma)(t\gamma)(t\beta) \geqq (s\beta)(t\beta)$ by TB4 and 13(a). Now assume that $(s\alpha)(t\alpha) \geqq (s\beta)(t\beta)$. By the regularity of $\mathcal{S}$, there is some δ in $\mathcal{S}$ such that $\beta\delta\beta = \beta$. Let $\gamma = \delta\alpha$. Then $(s\beta)(s\gamma) = (s\beta)(s\delta)(s\alpha) = (s\beta)(s\delta)(s\alpha)(t\alpha)(s\alpha) \geqq (s\beta)(s\delta)(s\beta)(t\beta)(s\alpha) = (s\beta\delta\beta)(t\beta)(s\alpha) = (s\beta)(t\beta)(s\alpha) \geqq (s\alpha)$ by TB6, 13(g), our assumption and 13(a), TB6, $\beta\delta\beta = \beta$, and TB4 respectively. Hence, for each $x \in A$, $(s\beta)(s\gamma) - x \geqq (s\alpha) - x$. Then $-(s\beta)(s\gamma)x \geqq -(s\alpha)x$, by TB1, and hence $(s\beta)(s\gamma) \leqq (s\alpha)$, by TB0. Therefore $(s\beta)(s\gamma) = (s\alpha)$.

(b) By TB6 and 17(a).

(c) Assume that $(s\alpha) = (s\gamma)(s\beta)$. By 15(a) and 15(b), $(t\alpha) = (t\beta)(t\gamma)$. Hence $(t\alpha)(s\alpha) = (t\beta)(t\gamma)(s\gamma)(s\beta) \leqq (t\beta)(s\beta)$ by TB5 and 13(b). Now assume that $(t\alpha)(s\alpha) \leqq (t\beta)(s\beta)$. By the regularity of $\mathcal{S}$, there is some δ in $\mathcal{S}$ such that $\beta\delta\beta = \beta$. Let $\gamma = \alpha\delta$. Then $(s\gamma)(s\beta) = (s\alpha)(s\delta)(s\beta) = (s\alpha)(t\alpha)(s\alpha)(s\delta)(s\beta) \leqq (s\alpha)(t\beta)(s\beta)(s\delta)(s\beta) = (s\alpha)(t\beta)(s\beta\delta\beta) = (s\alpha)(t\beta)(s\beta) \leqq (s\alpha)$ by TB6, 13(g), our assumption and 13(a), TB6, $\beta\delta\beta = \beta$, and TB5 and 13(a) respectively. By TB1 and TB0 it follows that $(s\gamma)(s\beta) = (s\alpha)$.

(d) By 17(c), 14(f), and TB0.

(e) By TB6 and 17(d).

(f) Assume that $\beta \leqq_R \gamma\alpha$. Then there is some δ such that $\delta\beta = \gamma\alpha$. Then $(t\gamma)\mathfrak{1} = (t\gamma)(s\delta)\mathfrak{1} \leqq (s\alpha)(t\beta)\mathfrak{1}$ by 13(d) and 16(b) respectively. ♦

Some consequences of TB0, ···, TB6, TB9 will be considered next. Part (a) of the following lemma expresses the reflexivity of equality, and corresponds to (C_5) of [5] and also to an axiom on p. 216 of [4]. To axioms (P_1) and (P_3) on p. 111 of [4] there correspond equalities in part (b). Part (f) complements TB10 and, in view of 14(h), may be used in place of TB9.

Lemma 18: *Assume that* $\mathfrak{A}$ *satisfies* TB0, ···, TB6, TB9. *For parts* (a), ···, (d) *assume also that* $\mathcal{S}$ *has an identity* id.

(a) $(t\,\mathrm{id})\mathfrak{1} = \mathfrak{1}$.

(b) $x = (s\,\mathrm{id})x = (t\,\mathrm{id})x = (s\,\mathrm{id})(t\,\mathrm{id})x$.

(c) $(t\alpha)x = (s\gamma)(s\alpha)(t\alpha)x$, *provided* $\gamma\alpha = \mathrm{id}$.

(d) $(t\alpha)x = (t\alpha)\mathfrak{1} \cdot (s\gamma)x$, *provided* $\alpha\gamma = \mathrm{id}$.

(e) $(s\alpha)(t\alpha)(t\beta)(s\beta) = (t\beta)(s\beta)(s\alpha)(t\alpha)$, *provided* β *is an* $\leqq_R$ *lower bound of* $\{\delta\colon \beta\alpha \leqq_R \delta\alpha\}$.

(f) $-(s\alpha)(t\alpha)-(t\beta)\mathfrak{1} = (t\beta)\mathfrak{1}$, *provided* β *is an* $\leqq_R$ *lower bound of* $\{\delta\colon \beta\alpha \leqq_R \delta\alpha\}$.

Proof:

(a) $(t\,\mathrm{id})\mathfrak{1} = (s\,\mathrm{id})(t\,\mathrm{id})(t\,\mathrm{id})\mathfrak{1} = (s\,\mathrm{id})(t\,\mathrm{id})\mathfrak{1} = \mathfrak{1}$ by TB9, by 16(a), and by TB4 and TB0 respectively.

(b) $(s\,\mathrm{id})x = (s\,\mathrm{id})(t\,\mathrm{id})(s\,\mathrm{id})x = (s\,\mathrm{id})(t\,\mathrm{id})(t\,\mathrm{id})(s\,\mathrm{id})x =$

$(s\,\mathrm{id})(t\,\mathrm{id})((t\,\mathrm{id})\mathfrak{1} \cdot x) = (s\,\mathrm{id})(t\,\mathrm{id})(\mathfrak{1} \cdot x) = (s\,\mathrm{id})(t\,\mathrm{id})x$ by 13(g), 16(a), 14(f), 17(a), and TB0 respectively. Similarly $(t\,\mathrm{id})x = (s\,\mathrm{id})(t\,\mathrm{id})x$ by 13(h), TB6, 14(f), 18(a), and TB0 respectively. Hence $(s\,\mathrm{id})x = (t\,\mathrm{id})x$. It follows that $(s\,\mathrm{id})x = (s\,\mathrm{id})(s\,\mathrm{id})x = (s\,\mathrm{id})(t\,\mathrm{id})x \geqq x$ by TB4. But also $(s\,\mathrm{id})x = (s\,\mathrm{id})(s\,\mathrm{id})x = (t\,\mathrm{id})(s\,\mathrm{id})x \leqq x$ by TB5.

(c) $(t\alpha) = (s\,\mathrm{id})(t\alpha) = (s\gamma\alpha)(t\alpha) = (s\gamma)(s\alpha)(t\alpha)$ by 18(b), $\gamma\alpha = \mathrm{id}$, and TB6 respectively.

(d) $(t\alpha)x = (t\alpha)(s\,\mathrm{id})x = (t\alpha)(s\alpha\gamma)x = (t\alpha)(s\alpha)(s\gamma)x = (t\alpha)\mathfrak{1} \cdot (s\gamma)x$ by 18(b), $\alpha\gamma = \mathrm{id}$, TB6, and 14(f) respectively.

(e) Assume that β is an $\leqq_R$ lower bound of $\{\delta\colon \beta\alpha \leqq_R \delta\alpha\}$. Then $(s\alpha)(t\alpha)(t\beta)(s\beta)x = (s\alpha)(t\alpha)((t\beta)\mathfrak{1} \cdot x) = (s\alpha)(t\alpha)((s\alpha)(t\alpha)(t\beta)\mathfrak{1} \cdot x) = (s\alpha)(t\alpha)(t\beta)\mathfrak{1} \cdot (s\alpha)(t\alpha)x = (t\beta)\mathfrak{1} \cdot (s\alpha)(t\alpha)x = (t\beta)(s\beta)(s\alpha)(t\alpha)x$ by 14(f), TB9, 14(e), TB9, and 14(f) respectively.

(f) By TB9 and 14(h). ♦

We now turn briefly to TB0, $\cdots$, TB6, TB8, TB9. Part (a) of the following lemma is related to (D1) on p. 260 of [9]. Note that if $\mathcal{S}$ is one of the semigroups of Theorem 3 then, by Lemma 7, $\mathcal{S}$ contains for any elements α and β an element γ satisfying (i) and (ii).

LEMMA 19: *Assume that* $\mathfrak{A}$ *satisfies* TB0, $\cdots$, TB6, TB8, TB9 *and that* $\mathcal{S}$ *is regular.*

(a) $(s\alpha)(t\beta)\mathfrak{1} = (t\gamma)\mathfrak{1}$, *provided* (i) γ *is an* $\leqq_R$ *lower bound of* $\{\delta\colon \gamma\alpha \leqq_R \delta\alpha\}$, *and* (ii) $\gamma\alpha$ *is an* $\leqq_R$ lub of $\{\alpha, \beta\}$.

(b) $(s\alpha)(t\beta)(s\beta) = (t\gamma)(s\gamma)(s\alpha)$, *provided* α, β, γ *satisfy* (i) *and* (ii) *above.*

Proof:

(a) From (ii), which implies $\beta \leqq_R \gamma\alpha$, and from 17(f) it follows that $(s\alpha)(t\beta)\mathfrak{1} \geqq (t\gamma)\mathfrak{1}$. By 14(f), by (ii) and TB8, and by 16(a) respectively, $(t\alpha)(s\alpha)(t\beta)\mathfrak{1} = (t\alpha)\mathfrak{1} \cdot (t\beta)\mathfrak{1} \leqq (t\gamma\alpha)\mathfrak{1} = (t\alpha)(t\gamma)\mathfrak{1}$. Hence, by TB0, $(t\alpha)(s\alpha)(t\beta)\mathfrak{1} \cdot -(t\alpha)(t\gamma)\mathfrak{1} = \varnothing$. By 14(c), by TB1, by (i) and TB9, and by TB0 respectively, this is equivalent to each of the following conditions: $(s\alpha)(t\beta)\mathfrak{1} \cdot (s\alpha) - (t\alpha)(t\gamma)\mathfrak{1} = \emptyset$; $(s\alpha)(t\beta)\mathfrak{1} \cdot -(s\alpha)(t\alpha)(t\gamma)\mathfrak{1} = \emptyset$; $(s\alpha)(t\beta)\mathfrak{1} \cdot -(t\gamma)\mathfrak{1} = \emptyset$; $(s\alpha)(t\beta)\mathfrak{1} \leqq (t\gamma)\mathfrak{1}$.

(b) By 14(f), 13(e), 19(a), and 14(f). ♦

We now turn to TB0, ···, TB7, TB9. Parts (b) and (c) of the following lemma are related to (P_4) and (P_6) respectively on pp. 111–112 of [4]. Part (c) may also be compared and contrasted with Lemma 19(b) above. Note that if $\mathcal{S}$ has an identity then $\alpha\gamma$ is an $\leqq_L$ lub of $\{\alpha, \beta\}$ for $\beta = \alpha\gamma$.

LEMMA 20: *Assume that* $\mathfrak{A}$ *satisfies* TB0, ···, TB7, TB9 *and that* $\mathcal{S}$ *is regular. For parts* (b) *and* (c) *assume also that* $\mathcal{S}$ *satisfies* S6, S7, S8.

(a) $(s\alpha)(t\beta) = (t\beta)(s\alpha)$, *provided* (i) α *and* β *are idempotent,* (ii) β *is an* $\leqq_R$ *lower bound of* $\{\delta: \beta\alpha \leqq_R \delta\alpha\}$, *and* (iii) α *is an* $\leqq_R$ *lower bound of* $\{\delta: \alpha\beta \leqq_R \delta\beta\}$.

(b) $(s\alpha)(t\alpha)(s\beta)(t\beta) = (s\gamma)(t\gamma)$, *provided* γ *is an* $\leqq_L$ lub *of* $\{\alpha, \beta\}$.

(c) $(s\beta)(t\beta)(s\alpha) = (s\alpha)(s\gamma)(t\gamma)$, *provided* (i) α *is an* $\leqq_R$ *lower bound of* $\{\delta: \alpha\gamma \leqq_R \delta\gamma\}$ *and* (ii) $\alpha\gamma$ *is an* $\leqq_L$ lub *of* $\{\alpha, \beta\}$.

Proof:

(a) Assume (i), (ii), and (iii). Then

$$\begin{aligned}(s\alpha)(t\beta) &= (s\alpha)(t\alpha)(s\alpha)(t\beta)(s\beta)(t\beta)\\ &= (s\alpha)(t\alpha)(t\alpha)(s\alpha)(t\beta)(s\beta)(s\beta)(t\beta)\\ &= (t\beta)(s\beta)(s\alpha)(t\alpha)(t\alpha)(s\alpha)(s\beta)(t\beta)\\ &= (t\beta)(s\beta)(s\beta)(t\beta)(s\alpha)(t\alpha)(t\alpha)(s\alpha)\\ &= (t\beta)(s\beta)(t\beta)(s\alpha)(t\alpha)(s\alpha)\\ &= (t\beta)(s\alpha)\end{aligned}$$

by 13(g) and 13(h), by (i), 16(a), and TB6, by 14(g),

(ii), and 18(e), by (iii), 18(e), and TB7, by (i), TB6, and 16(a), and by 13(g) and 13(h) respectively.

(b) Assume that γ is an $\leqq_L$ lub of $\{\alpha, \beta\}$. Consider first the case where id is an $\leqq$ glb of $\{\alpha, \beta\}$. By 17(b) and S8, we can assume that (i) α and β are idempotent, and (iv) $\alpha\beta = \beta\alpha$. Since $\alpha \leqq_L \alpha\beta = \beta\alpha$, it follows from S6 that (ii) β is an $\leqq_R$ lower bound of $\{\delta\colon \beta\alpha \leqq_R \delta\alpha\}$. Since $\beta \leqq_L \beta\alpha = \alpha\beta$, it likewise follows that (iii) α is an $\leqq_R$ lower bound of $\{\delta\colon \alpha\beta \leqq_R \delta\beta\}$. Since $\alpha\beta$ is an $\leqq_L$ upper bound of $\{\alpha, \beta\}$, therefore $\gamma \leqq_L \alpha\beta$. By $5(\mathrm{a})_L$, $\alpha\gamma = \gamma$ and $\beta\gamma = \gamma$. Hence $\gamma = \alpha\gamma = \alpha\beta\gamma$, so that $\alpha\beta \leqq_L \gamma$. Thus $\alpha\beta =_L \gamma$. It follows that $(s\gamma)(t\gamma) = (s\alpha\beta)(t\alpha\beta) = (s\alpha\beta)(t\beta\alpha) = (s\alpha)(s\beta)(t\alpha)(t\beta) = (s\alpha)(t\alpha)(s\beta)(t\beta)$ by 17(b), $\alpha\beta = \beta\alpha$, TB6 and 16(a), and 20(a) respectively.

Now consider the case where id is not an $\leqq_L$ glb of $\{\alpha, \beta\}$. By S7, there are α' and β' such that $\alpha' \leqq_L \alpha$, $\beta' \leqq_L \beta$, γ is an $\leqq_L$ lub of $\{\alpha, \beta\}$, and id is an $\leqq_L$ glb of $\{\alpha', \beta'\}$. By the previous case, $(s\gamma)(t\gamma) = (s\alpha')(t\alpha')(s\beta')(t\beta')$. From 17(b) together with 13(a), 13(b), and TB0 it follows that $(s\alpha')(t\alpha')(s\beta')(t\beta') \leqq (s\alpha)(t\alpha)(s\beta)(t\beta)$ and that $(s\alpha)(t\alpha)(s\beta)(t\beta) \leqq (s\gamma)(t\gamma)(s\gamma)(t\gamma)$. From 13(g) and TB0 it follows that $(s\alpha)(t\alpha)(s\beta)(t\beta) = (s\gamma)(t\gamma)$.

(c) Assume (i) and (ii). Then

$$\begin{aligned}(t\alpha)(s\beta)(t\beta)(s\alpha) &= (t\alpha)(s\alpha)(t\alpha)(s\beta)(t\beta)(s\alpha)\\ &= (t\alpha)(s\alpha\gamma)(t\alpha\gamma)(s\alpha)\\ &= (t\alpha)(s\alpha)(s\gamma)(t\gamma)(t\alpha)(s\alpha)\\ &= (t\alpha)(s\alpha)(t\alpha)(s\alpha)(s\gamma)(t\gamma)\\ &= (t\alpha)(s\alpha)(s\gamma)(t\gamma)\end{aligned}$$

by 13(h), (ii) and 20(b), TB6 and 16(a), (i) and 18(e), and 13(g) respectively. It follows that $(s\beta)(t\beta)(s\alpha) = (s\beta)(t\beta)(s\alpha)(t\alpha)(s\alpha) = (s\alpha)(t\alpha)(s\beta)(t\beta)(s\alpha) = (s\alpha)(t\alpha)(s\alpha)(s\gamma)(t\gamma) = (s\alpha)(s\gamma)(t\gamma)$ by 13(g), TB7, the identity just derived, and 13(g) respectively. ♦

4. EQUIVALENCE OF TWO THEORIES

In §4 we shall show that, for many $\mathcal{S}$, the theory determined by $TB_{\mathcal{S}}^{-}$, i.e., by TB0, $\cdots$, TB9, is strongly equivalent to the theory, adjusted to $\mathcal{S}$, of polyadic algebras with diagonal elements. The equivalence is strong because it is based on definitions and because these definitions are equational and essentially unique.

In §4, $\mathcal{S}$ shall be a semigroup of transformations on I which has an identity id. Also (in contrast to §1) $J, K, \cdots$ shall be elements of $\mathcal{C}(\mathcal{S})$, and $M, N, \cdots$ shall be elements of $\mathcal{E}(\mathcal{S})$. Otherwise, the conventions of §3 shall remain in force.

We begin with two lemmas about $\mathcal{S}$. If $\alpha(i) \neq \alpha(k)$ whenever i and k are in J and $i \neq k$, then α shall be *one-to-one on J.*

LEMMA 21: *Assume that $\mathcal{S}$ is regular.*

(a) *If* ran $\alpha \cdot$ ran β = ran γ, *then γ is an $\leqq_L$* lub *of* $\{\alpha, \beta\}$.

(b) *If* ran $\gamma = \alpha^{-1}$(ran β) *and α is one-to-one on* $-$ran γ, *then* (i) *α is an $\leqq_R$ lower bound of* $\{\delta: \alpha\gamma \leqq_R \delta\gamma\}$ *and* (ii) *$\alpha\gamma$ is an $\leqq_L$* lub *of* $\{\alpha, \beta\}$.

(c) *If* $\gamma^{\smallsmile}\gamma$ = eqv $\alpha^*(\beta^{\smallsmile}\beta)$, *then* (i) *$\gamma$ is an $\leqq_R$ lower bound of* $\{\delta: \gamma\alpha \leqq_R \delta\alpha\}$ *and* (ii) *$\gamma\alpha$ is an $\leqq_R$* lub *of* $\{\alpha, \beta\}$.

Proof:

(a) By 6(b) and the last part of the proof of Theorem 10.

(b) Assume that ran $\gamma = \alpha^{-1}$(ran β) and that α is one-to-one on $-$ran γ. Consider any $i \in -$ran γ and any $k \neq i$. If $k \in -$ran γ, then $\alpha(i) \neq \alpha(k)$ since α is one-to-one on $-$ran γ. If $k \in$ ran γ, then $\alpha(k) \in$ ran β while $\alpha(i) \in -$ran β, so that again $\alpha(i) \neq \alpha(k)$. It follows that $(\alpha^{\smallsmile}\alpha) \upharpoonright$ (ran γ) $= \alpha^{\smallsmile}\alpha$. Then, by the first part of the proof of 6(d), α is an $\leqq_R$ lower bound of $\{\delta: \alpha\gamma \leqq_R \delta\gamma\}$. From ran $\gamma = \alpha^{-1}$(ran β) it follows that ran $(\alpha\gamma)$ = ran $\alpha \cdot$ ran β. Then, by 21(a), $\alpha\gamma$ is an $\leqq_L$ lub of $\{\alpha, \beta\}$.

(c) Assume that $\gamma^{\smallsmile}\gamma$ = eqv $\alpha^*(\beta^{\smallsmile}\beta)$. Then $(\gamma^{\smallsmile}\gamma) \upharpoonright$ (ran α) $= \gamma^{\smallsmile}\gamma$. Hence, by the first part of the proof of 6(d), γ is an $\leqq_R$ lower bound of $\{\delta: \gamma\alpha \leqq_R \delta\alpha\}$. Also, by the last part of the proof of Lemma 7, $(\gamma\alpha)\check{}(\gamma\alpha) = \text{eqv}(\alpha\check{}\alpha + \beta\check{}\beta)$. Hence, by the first part of the proof of 6(c), $\gamma\alpha$ is an $\leqq_R$ lub of $\{\alpha, \beta\}$. ♦

LEMMA 22:

(a) *Assume that* $\mathcal{S}$ *satisfies* S_t2. *Then for each* M *there is some* γ *such that* $M = \gamma\check{}\gamma$.

(b) *Assume that* $\mathcal{S}$ *satisfies* E_L2, E_L5, *and* S8. *Then for each* $J \neq I$ *there is some* γ *such that* $J = -\text{ran}\ \gamma$.

(c) *Assume that* $\mathcal{S}$ *satisfies* E_L2, E_L5, S1, S_t3, S7, *and* S8. *Then* $I \in \mathcal{C}(\mathcal{S})$ *implies that, for each* k, *there are* γ *and* δ *such that* $\{k\} = -\text{ran}\ \gamma$ *and* $-\{k\} = -\text{ran}\ \delta$.

Proof:

(b) Consider any α and β such that $\text{ran}\ \alpha \cdot \text{ran}\ \beta \neq \emptyset$. By E_L5, $\{\alpha, \beta\}$ has an $\leqq_L$ upper bound. By S8, there are idempotent α', β' such that $\alpha' =_L \alpha$, $\beta' =_L \beta$, and $\alpha'\beta' = \beta'\alpha'$. Then $\alpha' \leqq_L \alpha'\beta'$ and $\beta' \leqq_L \beta'\alpha' = \alpha'\beta'$. Now suppose that $\alpha' \leqq_L \gamma$ and $\beta' \leqq_L \gamma$. By $5(a)_L$, $\alpha'\gamma = \gamma$ and $\beta'\gamma = \gamma$. Hence $\alpha'\beta'\gamma = \alpha'\gamma = \gamma$, so that $\alpha'\beta' \leqq_L \gamma$. It follows that $\alpha'\beta'$ is an $\leqq_L$ lub of $\{\alpha', \beta'\}$ and hence of $\{\alpha, \beta\}$. By E_L2, $\text{ran}(\alpha'\beta') = \text{ran}\ \alpha \cdot \text{ran}\ \beta$. By induction, if $J = -(\text{ran}\ \alpha_0) + \cdots + -(\text{ran}\ \alpha_n)$ and $J \neq I$, then there is some γ such that $J = -\text{ran}\ \gamma$.

(c) By S_t3, each $\{k\}$ is $-\text{ran}\ \gamma$ for some γ. Now assume that I is in $\mathcal{C}(\mathcal{S})$. By 22(b), there are α' and β' such that $(-\text{ran}\ \alpha') + (-\text{ran}\ \beta') = I$. Then $\text{ran}\ \alpha' \cdot \text{ran}\ \beta' = \emptyset$ and hence $\{\alpha', \beta'\}$ has no $\leqq_L$ upper bound. By S7, there are α and β such that $\{\alpha, \beta\}$ has no $\leqq_L$ upper bound and id is an $\leqq_L$ glb of $\{\alpha, \beta\}$. By E_L5, $\text{ran}\ \alpha \cdot \text{ran}\ \beta = \emptyset$. Now $\text{ran}\ \alpha + \text{ran}\ \beta \neq I$ would imply that $\text{ran}\ \alpha + \text{ran}\ \beta \subseteq \text{ran}(i/j)$ where $i \in -\text{ran}\ \alpha \cdot -\text{ran}\ \beta$ and $j \in \text{ran}\ \alpha$, and hence, by 6(b), that $(i/j) \leqq_L \alpha$ and $(i/j) \leqq_L \beta$, so that id would not be an $\leqq_L$ glb of $\{\alpha, \beta\}$. It follows that $\text{ran}\ \alpha + \text{ran}\ \beta = I$. Now consider any k. Suppose $k \in \text{ran}\ \alpha$. Let $i \in \text{ran}\ \beta$. Then $\{k\} = \text{ran}\ \alpha \cdot \text{ran}((i/k) \circ \beta)$. Now suppose that $k \in -\text{ran}\ \alpha$. Let $i \in \text{ran}\ \alpha$. Then $\{k\} = \text{ran}\ \beta \cdot \text{ran}((i/k) \circ \alpha)$. In either case, by 22(b), $-\{k\}$ is $-\text{ran}\ \delta$ for some δ. ♦

In order to formulate certain definitions, we now select representatives from certain classes. From I we select an element i_0. For any M which is $\gamma\check{}\gamma$ for some γ, we select one such γ. For any J which is $-\text{ran}\,\gamma$ for some γ, we select one such γ. For any α in $-\mathcal{I}(\mathcal{S}) \cdot \mathcal{R}(\mathcal{S})$ we select one γ such that $\gamma\alpha = \text{id}$. For any α in $-\mathcal{I}(\mathcal{S}) \cdot -\mathcal{R}(\mathcal{S}) \cdot \mathcal{L}(\mathcal{S}) = -\mathcal{R}(\mathcal{S}) \cdot \mathcal{L}(\mathcal{S})$ we select one γ such that $\alpha\gamma = \text{id}$. Finally, for any α in

$$-\mathcal{I}(\mathcal{S}) \cdot -\mathcal{R}(\mathcal{S}) \cdot -\mathcal{L}(\mathcal{S}) \cdot \text{cl}(\mathcal{I}(\mathcal{S}) + \mathcal{R}(\mathcal{S}) + \mathcal{L}(\mathcal{S}))$$

we select one $\langle\gamma_0, \gamma_1, \cdots, \gamma_n\rangle$ such that α is $\gamma_0\gamma_1\cdots\gamma_n$ and each γ_i is in $\mathcal{I}(\mathcal{S}) + \mathcal{R}(\mathcal{S}) + \mathcal{L}(\mathcal{S})$.

Below, we list conditions on an algebra

$$\mathfrak{A} = \langle A, +, \cdot, -, \mathit{1}, (s\alpha), (t\alpha), (cJ), (dM)\rangle_{\alpha\in\mathcal{S}, J\in\mathcal{C}(\mathcal{S}), M\in\mathcal{E}(\mathcal{S})}$$

where each (cJ) is a 1-ary operation on A and each (dM) is a 0-ary operation on A. One may regard D1, D2, D3 as definitions of certain (dM) and (cJ) in terms of $\mathit{1}$, $(s\gamma)$, and $(t\gamma)$. By Lemma 22, if $\mathcal{S}$ satisfies $\text{E}_\text{L}2$, $\text{E}_\text{L}5$, S1, $\text{S}_\text{t}2$, $\text{S}_\text{t}3$, S7, S8, then all (cJ) and (dM) are thus defined. Likewise, one may regard D4, $\cdots$, D7 as definitions of certain $(t\alpha)$ in terms of $\cdot$, $(d\alpha\check{}\alpha)$, $(c - \text{ran}\,\alpha)$, and $(s\gamma)$. If $\mathcal{S}$ satisfies S0, then all $(t\alpha)$ are thus defined.

D1. $(dM) = (t\gamma)\mathit{1}$, provided γ is selected for M.

D2. $(cJ)x = (s\gamma)(t\gamma)x$, provided γ is selected for J.

D3. $(cI)x = (c\{i_0\})(c - \{i_0\})x$, provided $\{i_0\}$ and $-\{i_0\}$ are in $\mathcal{C}(\mathcal{S})$.

D4. $(t\alpha)x = (d\alpha\check{}\alpha) \cdot (c - \text{ran}\,\alpha)x$, provided $\alpha \in \mathcal{I}(\mathcal{S})$.

D5. $(t\alpha)x = (s\gamma)(c - \text{ran}\,\alpha)x$, provided $\alpha \in -\mathcal{I}(\mathcal{S}) \cdot \mathcal{R}(\mathcal{S})$ and γ is selected for α.

D6. $(t\alpha)x = (d\alpha\check{}\alpha) \cdot (s\gamma)x$, provided $\alpha \in -\mathcal{R}(\mathcal{S}) \cdot \mathcal{L}(\mathcal{S})$ and γ is selected for α.

D7. $(t\alpha)x = (t\gamma_n) \cdots (t\gamma_1)(t\gamma_0)x$, provided

$$\alpha \in -\mathcal{I}(\mathcal{S}) \cdot -\mathcal{R}(\mathcal{S}) \cdot -\mathcal{L}(\mathcal{S})$$

and $\langle\gamma_0, \gamma_1, \cdots, \gamma_n\rangle$ is selected for α.

We now list conditions on an algebra

$$\mathfrak{A} = \langle A, +, \cdot, -, \mathit{1}, (s\alpha), (cJ), (dM)\rangle_{\alpha\in\mathcal{S}, J\in\mathcal{C}(\mathcal{S}), M\in\mathcal{E}(\mathcal{S})},$$

where each (cJ) is a 1-ary operation on A and each (dM) is a 0-ary operation on A. We regard each of PD0, ⋯, PD15 as a set, and we let $\mathrm{PD}_{\mathcal{S}}$ be their union. Then PD0, ⋯, PD11 consist of the axioms, adjusted to $\mathcal{S}$, for polyadic algebras, as given by Halmos [4], pp. 111–112, while PD12, ⋯, PD15 consist of the axioms, adjusted to $\mathcal{S}$, for diagonal elements, as given by Lucas [9], p. 260. Thus, $\mathrm{PD}_{\mathcal{S}}$ is the set of axioms, adjusted to $\mathcal{S}$, for polyadic algebras with diagonal elements. This adjustment requires (as we do) that $\mathcal{S}$ has an identity but neither that $J \in \mathcal{C}(\mathcal{S})$ implies $\alpha^{-1}J \in \mathcal{C}(\mathcal{S})$ nor that $M \in \mathcal{E}(\mathcal{S})$ implies $\mathrm{eqv}(\alpha^* M) \in \mathcal{E}(\mathcal{S})$.

PD0. $\langle A, +, \cdot, -, \mathcal{I}\rangle$ is a Boolean algebra.
PD1. $(s\alpha) - x = -(s\alpha)x$.
PD2. $(s\alpha)(x + y) = (s\alpha)x + (s\alpha)y$.
PD3. $(s\,\mathrm{id})x = x$.
PD4. $(s\alpha)(s\beta)x = (s\alpha\beta)x$.
PD5. $(cJ)\varnothing = \varnothing$.
PD6. $(cJ)x \geqq x$.
PD7. $(cJ)(x \cdot (cJ)y) = (cJ)x \cdot (cJ)y$.
PD8. $(c\varnothing)x = x$.
PD9. $(cJ)(cK)x = (c(J + K))x$.
PD10. $(s\alpha)(cJ)x = (s\beta)(cJ)x$, provided α and β agree on $-J$.
PD11. $(cJ)(s\alpha)x = (s\alpha)(cK)x$, provided $K = \alpha^{-1}J$ and α is one-to-one on K.
PD12. $(s\alpha)(dM) = (dN)$, provided $N = \mathrm{eqv}(\alpha^* M)$.
PD13. $(d\,\mathrm{id}) = \mathcal{I}$.
PD14. $x \cdot (d\alpha\breve{}\alpha) \leqq (s\alpha)x$, provided $\alpha\alpha = \alpha$.
PD15. $(dM) \cdot (dN) \leqq (d\,\mathrm{eqv}(M + N))$.

We now, roughly speaking, interpret the theory determined by $\mathrm{PD}_{\mathcal{S}}$ in the theory determined by $\mathrm{TB}_{\mathcal{S}}^{-}$.

THEOREM 23: *Assume that* $\mathcal{S}$ *satisfies* $\mathrm{E_L}2$, $\mathrm{E_L}5$, S1, $\mathrm{S_t}2$, S6, S7, S8 *and, if* $I \in \mathcal{C}(\mathcal{S})$, also $\mathrm{S_t}3$. *Further assume that* $\mathfrak{A}$ *satisfies* TB0, ⋯, TB9 *and* D1, D2, D3. *Then* $\mathfrak{A}$ *satisfies* PD0, ⋯, PD15 *and* D4, ⋯, D7.

Proof: From D1, D2, D3 and from Lemma 22 it follows that each (dM) is some $(t\gamma)\mathcal{I}$ such that $\gamma\breve{}\gamma = M$, each (cJ) other than

(cI) is some $(s\gamma)(t\gamma)$ such that $J = -\text{ran}\,\gamma$ and, if $I \in \mathfrak{C}(\mathcal{S})$, then $(cI) = (c\{i_0\})(c - \{i_0\})$ where $\{i_0\} \neq I$ and $-\{i_0\} \neq I$. In the rest of the proof we shall use these facts tacitly.

Since $\mathfrak{A}$ satisfies TB0, TB1, TB2, TB6, therefore $\mathfrak{A}$ satisfies PD0, PD1, PD2, PD4. By S7 and 18(b), $\mathfrak{A}$ satisfies PD3. By 13(i), $\mathfrak{A}$ satisfies PD5. By TB4, $\mathfrak{A}$ satisfies PD6. By 14(e) and, if $J = I$, also TB7, $\mathfrak{A}$ satisfies PD7. By S7, 18(b), and 17(b), $\mathfrak{A}$ satisfies PD8.

Consider any J and K. Consider first the case where $J + K \neq I$. Then, there are α, β, γ such that (cJ) is $(s\alpha)(t\alpha)$, (cK) is $(s\beta)(t\beta)$, $(c(J + K))$ is $(s\gamma)(t\gamma)$, and $\text{ran}\,\alpha \cdot \text{ran}\,\beta = \text{ran}\,\gamma$. By 20(b), $(s\alpha)(t\alpha)(s\beta)(t\beta) = (s\gamma)(t\gamma)$ and hence $(cJ)(cK) = (c(J + K))$. Consider now the case where $J + K = I$. Suppose first that $J = I$ and $K = I$. Then $(cJ)(cK)$ is $(c\{i_0\})(c - \{i_0\})(c\{i_0\})(c - \{i_0\})$. By TB7, this is $(c\{i_0\})(c\{i_0\})(c - \{i_0\})(c - \{i_0\})$. By the first case, this is $(c\{i_0\})(c - \{i_0\})$, hence (cI), and thus $(c(J + K))$. Suppose next that $J = I$ and $K \neq I$. By the first case, $(cK) = (c(K \cdot \{i_0\})(c(K \cdot -\{i_0\}))$. Then by TB7,

$$(cJ)(cK) = (c\{i_0\})(c - \{i_0\})(c(K \cdot \{i_0\}))(c(K \cdot -\{i_0\}))$$

$$= (c\{i_0\})(c(K \cdot \{i_0\})(c - \{i_0\})(c(K \cdot -\{i_0\})).$$

By the first case, this is $(c\{i_0\})(c - \{i_0\})$, hence (cI), and thus $(c(J + K))$. When $J \neq I$ and $K = I$ or when $J \neq I$ and $K \neq I$, then a similar proof shows that $(cJ)(cK) = (c(J + K))$, which therefore holds in all cases. Thus $\mathfrak{A}$ satisfies PD9.

Consider any α, β, J such that α and β agree on $-J$. Consider first the case where $J \neq I$. Then (cJ) is $(s\gamma)(t\gamma)$ for some γ such that $-J = \text{ran}\,\gamma$. Then $\alpha\gamma = \beta\gamma$ and hence, by TB6, $(s\alpha)(s\gamma) = (s\alpha\gamma) = (s\beta\gamma) = (s\beta)(s\gamma)$. Hence $(s\alpha)(cJ) = (s\alpha)(s\gamma)(t\gamma) = (s\beta)(s\gamma)(t\gamma) = (s\beta)(cJ)$. Now consider the case where $J = I$ and hence (cJ) is $(c\{i_0\})(c - \{i_0\})$. By TB7, this is also $(c - \{i_0\})(c\{i_0\})$. By $\mathrm{S_t3}$, $(i_0/\alpha(i_0))$ is in $\mathcal{S}$. Now α and $(i_0/\alpha(i_0))$ agree on $\{i_0\}$, and id and $(i_0/\alpha(i_0))$ agree on $-\{i_0\}$. Hence, by the first case, $(s\alpha)(c - \{i_0\}) = (s(i_0/\alpha(i_0))(c - \{i_0\})$ and $(s(i_0/\alpha(i_0)))(c\{i_0\}) = (s\text{ id})(c\{i_0\})$. It follows that $(s\alpha)(cI) =$

$(s\,\mathrm{id})(cI)$. Similarly, $(s\beta)(cI) = (s\,\mathrm{id})(cI)$. Hence when $J = I$ then again $(s\alpha)(cJ) = (s\beta)(cJ)$. Thus $\mathfrak{A}$ satisfies PD10.

Consider any α, J, and K such that $K = \alpha^{-1}J$ and α is one-to-one on K. Consider first the case where $K \neq I$. Then also $J \neq I$. By 21(b), there are β and γ such that (cJ) is $(s\beta)(t\beta)$, (cK) is $(s\gamma)(t\gamma)$, α is an $\leqq_R$ lower bound of $\{\delta\colon \alpha\gamma \leqq_R \delta\gamma\}$, and $\alpha\gamma$ is an $\leqq_L$ lub of $\{\alpha, \beta\}$. By 20(c), $(s\beta)(t\beta)(s\alpha) = (s\alpha)(s\gamma)(t\gamma)$ and hence $(cJ)(s\alpha) = (s\alpha)(cK)$. Now consider the case where $K = \alpha^{-1}J = I$ and hence $(cK) = (c\{i_0\})(c - \{i_0\})$. Let $\alpha(i_0) = j_0$. Then

$$\begin{aligned}(cJ)(s\alpha) &= (c\{j_0\})(c(J \cdot -\{j_0\}))(s\alpha)\\ &= (c\{j_0\})(s\alpha)(c - \{i_0\})\\ &= (s\alpha)(c\{i_0\})(c - \{i_0\}) = (s\alpha)(cK)\end{aligned}$$

by PD9, the previous case, the previous case, and PD9 respectively. Thus $\mathfrak{A}$ satisfies PD11.

By 21(c) and 19(a), $\mathfrak{A}$ satisfies PD12. By S7, 18(a), and 17(e), $\mathfrak{A}$ satisfies PD13. By 16(g) and 17(e), $\mathfrak{A}$ satisfies PD14. By S_t2, 6(c), and TB8, $\mathfrak{A}$ satisfies PD15.

Consider any α in $\mathcal{I}(\mathcal{S})$. By 16(f) and by 17(e) and 17(b) respectively, $(t\alpha)x = (t\alpha)1 \cdot (s\alpha)(t\alpha)x = (d\alpha\check{\ }\alpha) \cdot (c - \mathrm{ran}\,\alpha)x$. Hence $\mathfrak{A}$ satisfies D4. Now consider any α and γ such that $\gamma\alpha = \mathrm{id}$. By 18(c) and 17(b) respectively, $(t\alpha)x = (s\gamma)(s\alpha)(t\alpha)x = (s\gamma)(c - \mathrm{ran}\,\alpha)x$. Hence $\mathfrak{A}$ satisfies D5. Next consider any α and γ such that $\alpha\gamma = \mathrm{id}$. By 18(d) and 17(e) respectively, $(t\alpha)x = (t\alpha)1 \cdot (s\gamma)x = (d\alpha\check{\ }\alpha) \cdot (s\gamma)x$. Hence $\mathfrak{A}$ satisfies D6. Finally, by 16(a), $\mathfrak{A}$ satisfies D7. ♦

Some well-known consequences of $PD_{\mathcal{S}}$ will be useful.

LEMMA 24: *Assume that* $\mathfrak{A}$ *satisfies* PD0, ···, PD11 *with the possible exception of* PD4 *and* PD9.

(a) $(s\alpha)(x \cdot y) = (s\alpha)x \cdot (s\alpha)y$.
(b) $(cJ)x \cdot y = \varnothing$ *if and only if* $x \cdot (cJ)y = \varnothing$.
(c) $(cJ)(x + y) = (cJ)x + (cJ)y$.

(d) *If* $x \leqq y$ *then* $(cJ)x \leqq (cJ)y$.
(e) $(c - \text{ran}\,\alpha)(s\alpha)x = (s\alpha)x$.
(f) $(s\alpha)(c - \text{ran}\,\alpha)x = (c - \text{ran}\,\alpha)x$, *provided* $\alpha\alpha = \alpha$.

Proof:

(b) Assume that $(cJ)x \cdot y = \varnothing$. Then

$$x \cdot (cJ)y \leqq (cJ)(x \cdot (cJ)y) = (cJ)x \cdot (cJ)y$$
$$= (cJ)((cJ)x \cdot y) = (cJ)\varnothing = \varnothing$$

by PD6, PD7, PD7 and PD0, our assumption, and PD5 respectively. By PD0, if $(cJ)x \cdot y = \varnothing$ then $x \cdot (cJ)y = \varnothing$. Similarly, if $x \cdot (cJ)y = \varnothing$, then $(cJ)x \cdot y = \varnothing$.

(c) By 24(b) and 15(c).

(e) Since $\varnothing = \alpha^{-1}(-\text{ran}\,\alpha)$ and α is one-to-one on $\varnothing$, it follows from PD11 and PD8 respectively that $(c - \text{ran}\,\alpha)(s\alpha)x = (s\alpha)(c\varnothing)x = (s\alpha)x$.

(f) Assume that $\alpha\alpha = \alpha$. Then α and id agree on ran α. By PD10 and PD3 respectively, $(s\alpha)(c - \text{ran}\,\alpha)x = (s\,\text{id})(c - \text{ran}\,\alpha)x = (c - \text{ran}\,\alpha)x$. ♦

The next lemma elaborates Theorem 1.1 of Jurie [8]. In their present form, the lemma and its proof have been suggested to me by Charles Pinter. By 5(c) and 6(b), the lemma implies that if $\mathfrak{A}$ satisfies $\text{PD}_{\mathcal{S}}$ and if $\mathcal{S}$ is regular then each (cJ) can be defined by a simple first-order formula in terms of $+$ and operations $(s\alpha)$.

LEMMA 25: *Assume that* $\mathfrak{A}$ *satisfies* PD0, $\cdots$, PD11 *with the possible exception of* PD4. *Consider any* $J = (-\text{ran}\,\alpha_0) + \cdots + (-\text{ran}\,\alpha_n)$ *such that* $\alpha_0, \cdots, \alpha_n$ *are idempotent. Then* $(cJ)x$ *is the least element of* $\{y\colon y \in \text{ran}(s\alpha_0), \cdots, y \in \text{ran}(s\alpha_n)$, *and* $y \geqq x\}$.

Proof: By PD6, $(cJ)x \geqq x$. Given any $i \leqq n$, it follows from PD9 and 24(f) respectively that $(cJ)x = (c - \text{ran}\,\alpha_i)(cJ)x = (s\alpha_i)(c - \text{ran}\,\alpha_i)x$. Hence $(cJ)x$ is in the set $\{y\colon y \in \text{ran}(s\alpha_0), \cdots, y \in \text{ran}(s\alpha_n)$, and $y \geqq x\}$. Now consider any y in this set. By 24(e), given any $i \leqq n$, there is some z such that $y = (s\alpha_i)z = (c - \text{ran}\,\alpha_i)(s\alpha_i)z = (c - \text{ran}\,\alpha_i)y$. By PD9,

$$y = (c - \text{ran}\,\alpha_0) \cdots (c - \text{ran}\,\alpha_n)y = (cJ)y.$$

From $y \geqq x$ and 24(d), it follows that $(cJ)y \geqq (cJ)x$ and hence that $y = (cJ)y \geqq (cJ)x$. ♦

The following lemma contains some results of [4], pp. 215–218. According to $5(c)_R$, 6(a), and 22(a), part (c) implies that if $\mathfrak{A}$ satisfies $PD_{\mathcal{S}}$ and $\mathcal{S}$ satisfies S1 and S_t2, then each (eM) can be defined by a simple first-order formula in terms of $+$ and an operation $(s\alpha)$.

LEMMA 26: *Assume that* $\mathfrak{A}$ *satisfies* PD0, PD1, PD12, PD13, PD14.

(a) $(s\alpha)(d\alpha^{\smile}\alpha) = \mathit{1}$.
(b) $(s\alpha)x \cdot (d\alpha^{\smile}\alpha) \leqq x$, *provided* $\alpha\alpha = \alpha$.
(c) *If* α *is idempotent, then* $(d\alpha^{\smile}\alpha)$ *is the least element of* $\{y\colon (s\alpha)y = \mathit{1}\}$.

Proof:

(a) Since $\alpha^*(\alpha^{\smile}\alpha) = \alpha\alpha^{\smile}\alpha\alpha^{\smile} \subseteq \text{id}$, therefore $\text{eqv}\,\alpha^*(\alpha^{\smile}\alpha) = \text{id}$. Hence, by PD12 and PD13 respectively, $(s\alpha)(d\alpha^{\smile}\alpha) = (d\,\text{id}) = \mathit{1}$.
(b) We use the argument of [4], p. 215. Assume that $\alpha\alpha = \alpha$. Then $(s\alpha)x \cdot -x \cdot (d\alpha^{\smile}\alpha) \leqq (s\alpha)x \cdot (s\alpha) - x = (s\alpha)x \cdot -(s\alpha)x = \varnothing$ by PD14, PD1, and PD0 respectively. By PD0, $(s\alpha)x \cdot (d\alpha^{\smile}\alpha) \leqq x$.
(c) Assume that $\alpha\alpha = \alpha$. By 26(a), $(d\alpha^{\smile}\alpha)$ is in the set $\{y\colon (s\alpha)y = \mathit{1}\}$. Now consider any y in this set. Then by PD0 and 26(b), $(d\alpha^{\smile}\alpha) = \mathit{1} \cdot (d\alpha^{\smile}\alpha) = (s\alpha)y \cdot (d\alpha^{\smile}\alpha) \leqq y$. ♦

The next lemma implies that if $\mathcal{S} \subseteq \text{cl}(\mathcal{J}(\mathcal{S}) + \mathcal{R}(\mathcal{S}) + \mathcal{L}(\mathcal{S}))$, then each $(s\alpha)$ of each $\mathfrak{A}$ satisfying $PD_{\mathcal{S}}$ has a conjugate.

LEMMA 27: *Assume that* $\mathfrak{A}$ *satisfies* PD0, $\cdots$, PD14 *with the possible exception of* PD4 *and* PD9. *Assume also that* $\mathcal{S} \subseteq \text{cl}(\mathcal{J}(\mathcal{S}) + \mathcal{R}(\mathcal{S}) + \mathcal{L}(\mathcal{S}))$, *so that for each* α *there is some* $(t\alpha)$, *necessarily unique, in the set of operations defined by* D4, $\cdots$, D7. *Then, for each* α, $(s\alpha)$ *and* $(t\alpha)$ *are conjugate.*

Proof: By D7 and 15(a), it suffices to show that $(s\alpha)$ and $(t\alpha)$ are conjugate whenever α is in $\mathcal{J}(\mathcal{S}) + \mathcal{R}(\mathcal{S}) + \mathcal{L}(\mathcal{S})$. Consider any such α. First, note that $(t\alpha)(x + y) = (t\alpha)x + (t\alpha)y$. For

α in $\mathcal{I}(\mathcal{S})$, this follows from D4, 24(c), and PD0. For α in $-\mathcal{I}(\mathcal{S}) \cdot \mathcal{R}(\mathcal{S})$, this follows from D5, 24(c), and PD2. And for α in $-\mathcal{R}(\mathcal{S}) \cdot \mathcal{L}(\mathcal{S})$, this follows from D6, PD2, and PD0.

Also note that $(s\alpha)(t\alpha)x \geqq x$. For α in $\mathcal{I}(\mathcal{S})$, this follows from D4, 24(a), 26(a), 24(f), PD0, and PD6. For α in $-\mathcal{I}(\mathcal{S}) \cdot \mathcal{R}(\mathcal{S})$ this follows from D5, PD4, PD10 (and the agreement of $\alpha\gamma$ and $\gamma\alpha$ on ran α whenever $\gamma\alpha = \text{id}$ and hence $\alpha\gamma\alpha = \gamma\alpha\alpha$), PD11, PD3, and PD6. And for α in $-\mathcal{R}(\mathcal{S}) \cdot \mathcal{L}(\mathcal{S})$, this follows from D6, 24(a), 26(a), PD0, PD4 and PD3.

Further note that $(t\alpha)(s\alpha)x \leqq x$. For α in $\mathcal{I}(\mathcal{S})$, this follows from D4, 24(e), and 26(b). For α in $-\mathcal{I}(\mathcal{S}) \cdot \mathcal{R}(\mathcal{S})$, this follows from D5, 24(e), PD4, PD3, and PD0. And for α in $-\mathcal{R}(\mathcal{S}) \cdot \mathcal{L}(\mathcal{S})$, this follows from D6, PD4, 26(b), and the fact that $(\beta\alpha)\breve{}(\beta\alpha) = \breve{\alpha}\breve{\beta}\beta\alpha = \breve{\alpha}\alpha$ and $\beta\alpha\beta\alpha = \beta\alpha$ whenever $\alpha\beta = \text{id}$.

From the proof of 14(c), it now follows that $(s\alpha)$ and $(t\alpha)$ are conjugate. ♦

We now, roughly speaking, interpret the theory determined by $TB_{\mathcal{S}}^{-}$ in the theory determined by $PD_{\mathcal{S}}$.

THEOREM 28: *Assume that* $\mathcal{S}$ *satisfies* S0, S1, S_t2, S_t3, S_t4, *and that* $\mathfrak{A}$ *satisfies* PD0, $\cdots$, PD15 *and* D4, $\cdots$, D7. *Then* $\mathfrak{A}$ *satisfies* TB0, $\cdots$, TB9 *and* D1, D2, D3.

Proof: Since $\mathfrak{A}$ satisfies PD0, PD1, PD2, PD4, therefore $\mathfrak{A}$ satisfies TB0, TB1, TB2, TB6. By Lemma 27, for each α, $(s\alpha)$ and $(t\alpha)$ are conjugate. As we noted after the proof of Lemma 14 (in discussing alternate axiom sets), this implies that $\mathfrak{A}$ satisfies TB3, TB4, TB5. From their proofs it now follows that $\mathfrak{A}$ satisfies 17(b) and 17(e).

Consider any M and γ such that γ has been selected for M. By $5(c)_R$ and 6(a), $\mathcal{S}$ contains an idempotent α such that $\alpha =_R \gamma$ and $\breve{\alpha}\alpha = \breve{\gamma}\gamma = M$. Then $(t\gamma)\mathbf{1} = (t\alpha)\mathbf{1} = (d\breve{\alpha}\alpha)\cdot(c-\text{ran}\,\alpha)\mathbf{1} = (dM)\cdot(c-\text{ran}\,\alpha)\mathbf{1} = (dM)$ by 17(e), D4, $\breve{\alpha}\alpha = M$, and PD6 and PD0 respectively. Hence $\mathfrak{A}$ satisfies D1. Now consider any J and γ such that γ has been selected for J. By $5(c)_L$ and 6(b), $\mathcal{S}$ contains an idempotent α such that $\alpha =_L \gamma$ and ran α = ran γ = $-J$. Then $(s\gamma)(t\gamma)x = (s\alpha)(t\alpha)x = (s\alpha)((d\breve{\alpha}\alpha)\cdot(c-\text{ran}\,\alpha)x) =$

$(s\alpha)(d\alpha^{\smallsmile}\alpha) \cdot (s\alpha)(c - \operatorname{ran} \alpha)x = \mathcal{I} \cdot (s\alpha)(c - \operatorname{ran} \alpha)x = (c - \operatorname{ran} \alpha)x = (cJ)x$ by 17(b), D4, 24(a), PD12 and PD13, PD0 and 24(f), and $\operatorname{ran} \alpha = -J$ respectively. Hence $\mathfrak{A}$ satisfies D2. Also, if $\{i_0\}$ and $-\{i_0\}$ are in $\mathcal{C}(\mathcal{S})$, then I is in $\mathcal{C}(\mathcal{S})$ so that, by PD9, $(cI)x = c\{i_0\})(c - \{i_0\})x$. Hence $\mathfrak{A}$ satisfies D3.

Now consider any α and β. Suppose that γ and δ have been selected for $-\operatorname{ran} \alpha$ and $-\operatorname{ran} \beta$ respectively. Then $(s\alpha)(t\alpha)(s\beta)(t\beta) = (s\gamma)(t\gamma)(s\delta)(t\delta) = (c - \operatorname{ran} \alpha)(c - \operatorname{ran} \beta) = c(-(\operatorname{ran} \alpha) + -(\operatorname{ran} \beta)) = (c - \operatorname{ran} \beta)(c - \operatorname{ran} \alpha) = (s\delta)(t\delta)(s\gamma)(t\gamma) = (s\beta)(t\beta)(s\alpha)(t\alpha)$ by 17(b), D2, PD9, PD9, D2, and 17(b) respectively. Hence $\mathfrak{A}$ satisfies TB7. Next consider any α, β, γ such that γ is an $\leqq_R$ lub of $\{\alpha, \beta\}$. By 6(c), $\gamma^{\smallsmile}\gamma = \operatorname{eqv}(\alpha^{\smallsmile}\alpha + \beta^{\smallsmile}\beta)$. Suppose that α', β', γ' have been selected for $\alpha^{\smallsmile}\alpha$, $\beta^{\smallsmile}\beta$, and $\gamma^{\smallsmile}\gamma$ respectively. Then $(t\alpha)\mathcal{I} \cdot (t\beta)\mathcal{I} = (t\alpha')\mathcal{I} \cdot (t\beta')\mathcal{I} = (d\alpha'^{\smallsmile}\alpha') \cdot (d\beta'^{\smallsmile}\beta') \leqq (d\gamma'^{\smallsmile}\gamma') = (t\gamma')\mathcal{I} = (t\gamma)\mathcal{I}$ by 17(e), D1, PD15, D1, and 17(e) respectively. Hence $\mathfrak{A}$ satisfies TB8. Finally, consider any α and β such that β is an $\leqq_R$ lower bound of $\{\delta: \beta\alpha \leqq_R \delta\alpha\}$. By 6(d), $(\beta^{\smallsmile}\beta) \upharpoonright (\operatorname{ran} \alpha) = \beta^{\smallsmile}\beta$. By $5(\mathrm{c})_{\mathrm{L}}$ and 6(b), $\mathcal{S}$ contains an idempotent α'' such that $\alpha'' =_L \alpha$ and $\operatorname{ran} \alpha'' = \operatorname{ran} \alpha$. From $(\beta^{\smallsmile}\beta) \upharpoonright \operatorname{ran} \alpha'' = \beta^{\smallsmile}\beta$ and $\alpha''\alpha'' = \alpha''$ it follows that $\operatorname{eqv} \alpha''*(\beta^{\smallsmile}\beta) = \beta^{\smallsmile}\beta$. By PD12, $(s\alpha'')(d\beta^{\smallsmile}\beta) = (d\beta^{\smallsmile}\beta)$. Suppose that α' and β' have been selected for $-\operatorname{ran} \alpha$ and $\beta^{\smallsmile}\beta$ respectively. Then $(s\alpha)(t\alpha)(t\beta)\mathcal{I} = (s\alpha')(t\alpha')(t\beta')\mathcal{I} = (c - \operatorname{ran} \alpha'')(d\beta^{\smallsmile}\beta) = (c - \operatorname{ran} \alpha'')(s\alpha'')(d\beta^{\smallsmile}\beta) = (s\alpha'')(d\beta^{\smallsmile}\beta) = (d\beta^{\smallsmile}\beta) = (t\beta')\mathcal{I} = (t\beta)\mathcal{I}$ by 17(b) and 17(e), D2 and D1, the identity just derived, 24(e), the identity just derived, D1, and 17(e) respectively. Hence $\mathfrak{A}$ satisfies TB9. ♦

Consider two given theories such that there is a theory which can be obtained from each by adjoining suitable definitions. The two given theories shall be *equivalent by definitions.* If, moreover, there is only one theory which can thus be obtained from each and which involves no notions other than those involved by at least one of the two given theories, then they shall be *uniquely equivalent by definitions.*

We are now ready to prove the main result of § 4. By Lemma 7, it applies to each semigroup $\mathcal{S}$ of Theorem 3.

THEOREM 29: *Assume that* $\mathcal{S}$ *satisfies* E_L2, E_L5 *and each of* S0, ···, S8 *with the possible exception of* S5. *Then the theories determined by* $TB_{\mathcal{S}}^{-}$ *and* $PD_{\mathcal{S}}$ *respectively are uniquely equivalent by definitions. Moreover, all definitions can be chosen to be equational.*

Proof: Equivalence by definitions, and also the existence of equational definitions in all cases, follows from Theorems 23 and 28. Uniqueness follows from 14(c) and 15(b), and from the regularity of $\mathcal{S}$, 5(c), 6(a), 6(b), 25, and 26(c). ♦

5. REPRESENTATION AND SIMPLIFICATION

We now let $\mathcal{S}$ be again an arbitrary semigroup. One representation problem for $TB_{\mathcal{S}}$ is that of finding a class of concrete algebras $\mathfrak{B}$ such that an algebra $\mathfrak{A}$ satisfies $TB_{\mathcal{S}}$ if and only if $\mathfrak{A}$ is isomorphic to some $\mathfrak{B}$ in the class. Below, using basic results of Jónsson and Tarski in [7], we reduce this representation problem to one for semigroups. Analogous uses of [7] occur in 2.7 of [5]. Such reductions are not solutions. Nevertheless, they have heuristic value.

A second aim of § 5 is to give conditions which can be used in place of $TB_{\mathcal{S}}$ and which involve $\mathcal{S}$ in a simpler way.

Consider, for example, an equality of the form TB10. The *scope* of the first and second $-$ in it shall be $(s\alpha)(t\alpha) - (t\beta)\mathbf{1}$ and $(t\beta)\mathbf{1}$ respectively. Thus, no variable in an equality of the form TB10 is in the scope of an $-$.

We now state, in a form easy for us to use, part of the basic result of [7]. For proof one uses the basic theorems 2.10 and 3.10 of [7], and also 3.6 of [7] and 14(c) above, and 3.5 of [7] and 13(d) and 13(e) above.

THEOREM 30: *Assume that* $\mathfrak{A}$ *satisfies* TB0, ···, TB5. *Then there is a prerepresentation* $\hat{}$ *of* $\mathcal{S}$ *such that, if* $\mathfrak{B}$ *is the complete algebra induced by* $\hat{}$, *then* (a) $\mathfrak{A}$ *is isomorphic to a subalgebra of* $\mathfrak{B}$, *and* (b) *any equality satisfied by* $\mathfrak{A}$ *and having no variable in the scope of an* $-$ *is also satisfied by* $\mathfrak{B}$.

Since an equality is satisfied by $\mathfrak{A}$ if and only if it is satisfied by any $\mathfrak{A}'$ that is isomorphic to $\mathfrak{A}$, one may regard 30(a) as the converse of 11(a). Except that TB0, $\cdots$, TB5 differ from the equational axioms suggested in [7], 30(a) is contained in [7]. As one can see from the proof of the next theorem, using 30(b) as well as 30(a) one obtains a similar converse of any combination of parts of Theorem 11 which includes part (a). In two of these cases one can also bring Theorem 10 into play, making essential use of TB10.[5] One of these is the case where $\mathfrak{A}$ satisfies each of TB0, $\cdots$, TB10, and the other is the case where $\mathfrak{A}$ satisfies each of these except perhaps TB7. For the first of these we shall now show this explicitly.

THEOREM 31: *Assume that* $\mathcal{S}$ *has an identity and satisfies* S1, S2, S5. *Then* $\mathfrak{A}$ *satisfies* TB0, $\cdots$, TB10 *if and only if* $\mathfrak{A}$ *is isomorphic to an algebra that is induced by an antirepresentation* $\hat{}$ *of* $\mathcal{S}$ *such that either* $\hat{}$ *is trivial or else* $\hat{}$ *is* $\langle \leqq_R, \leqq_L \rangle$*-faithful and* $\{\hat{\alpha} : \alpha \in \mathcal{S}\}$ *has suitable ranges and commuting equivalences.*

Proof: Assume that $\mathfrak{A}$ satisfies TB0, $\cdots$, TB10. By Theorem 30, there is a prerepresentation $\hat{}$ of $\mathcal{S}$ such that, for the complete algebra $\mathfrak{B}$ induced by $\hat{}$, (a) and (b) of Theorem 30 hold. Let $\mathfrak{J} = \{\hat{\alpha} : \alpha \in \mathcal{S}\}$ and let $W = \text{dom}\,\hat{\alpha}$ for some α. Since $\mathfrak{A}$ satisfies TB6, $\cdots$, TB10, it follows from (b) that $\mathfrak{B}$ satisfies TB6, $\cdots$, TB10. Since $\mathfrak{B}$ satisfies TB6, therefore $\hat{\alpha}^{-1}(\hat{\beta}^{-1}x) = \widehat{\alpha\beta}^{-1}x$ for each $x \subseteq W$. Then, in particular, for any v and w in W, v is in $\hat{\alpha}^{-1}(\hat{\beta}^{-1}\{w\})$ if and only if v is in $\widehat{\alpha\beta}^{-1}\{w\}$. It follows that $\hat{\beta}(\hat{\alpha}(v)) = \widehat{\alpha\beta}(v)$ for any v in W. Hence $\hat{}$ satisfies T1, and is an antirepresentation of $\mathcal{S}$. Since $\mathfrak{B}$ satisfies TB7, therefore, in particular, for any v and w in V, v is in $\hat{\alpha}^{-1}(\hat{\alpha}^*(\hat{\beta}^{-1}(\hat{\beta}^*\{w\})))$ if and only if v is in $\hat{\beta}^{-1}(\hat{\beta}^*(\hat{\alpha}^{-1}(\hat{\alpha}^*\{w\})))$. It follows that $\hat{}$ satisfies T2, i.e., that $\mathfrak{J}$ has commuting equivalences. Since $\hat{\delta}^*\hat{\imath} = \text{ran}\,\hat{\delta}$ and since $\mathfrak{B}$ satisfies TB8 and TB9, therefore $\hat{}$ satisfies T3 and T4. Now suppose that β is not an $\leqq_R$ lower bound of $\{\delta : \beta\alpha \leqq_R \delta\alpha\}$, and consider any w in $-\hat{\alpha}^{-1}(\hat{\alpha}^*(-\text{ran}\,\hat{\beta}))$. Since $\mathfrak{B}$ satisfies TB10, therefore $\hat{\gamma}^{-1}(\{w\} \cdot -\hat{\alpha}^{-1}(\hat{\alpha}(-\text{ran}\,\hat{\beta}))) = \{w\} \cdot -\hat{\alpha}^{-1}(\hat{\alpha}(-\text{ran}\,\hat{\beta}))$.

[5] For use of TB10 in a completeness proof, and also for independence questions concerning TB10, see Ch. 9 of [3].

But $\{w\} \cdot -\hat{\alpha}^{-1}(\hat{\alpha}(-\text{ran } \hat{\beta})) = \{w\}$. Hence $\hat{\gamma}^{-1}\{w\} = \{w\}$. Therefore w is an isolated point for $\hat{\gamma}$. Since $\hat{\gamma}$ is arbitrary, w is in isl$\mathfrak{J}$. It follows that ^ satisfies T5. By one of the two conditionals contained in Theorem 10, either ^ is trivial or else ^ is $\langle \leqq_R, \leqq_L \rangle$-faithful and $\mathfrak{J}$ has suitable ranges. The converse follows from the other conditional in Theorem 10 and from Theorem 11. ♦

Conditions TB8, TB9, and TB10 on $\mathfrak{A}$ involve $\mathcal{S}$ in a fairly complicated way. In contrast, condition C1 below relates $\mathfrak{A}$ and $\mathcal{S}$ rather simply and conditions C0, C2, C3 concern only $\mathfrak{A}$. Again, the conditions apply to arbitrary elements α, β, γ of $\mathcal{S}$.

C0. $(s\alpha)$ is the identity transformation on A.
C1. $(t\alpha)\mathbf{1} \leqq (t\beta)\mathbf{1}$ if and only if $\beta \leqq_R \alpha$.
C2. There is some δ such that $(s\alpha)(t\beta)\mathbf{1} = (t\delta)\mathbf{1}$.
C3. If $-(s\alpha)(t\alpha) - (t\beta)\mathbf{1} < (t\beta)\mathbf{1}$, then $(s\gamma)x = x$ for every $x \leqq -(s\alpha)(t\alpha) - (t\beta)\mathbf{1}$.

C2 is simply the condition that $\{(t\beta)\mathbf{1}: \beta \in \mathcal{S}\}$ is closed under each $(s\alpha)$. If $\mathfrak{A}$ satisfies TB0, TB4, TB5, then C0 implies that also each $(t\alpha)$ is the identity transformation on A. In that case we shall say that $\mathfrak{A}$ is $\{s, t\}$*-trivial.* (The corresponding condition for cylindric algebras is given in 1.3.10 of [5], while a condition corresponding to C1 is involved in 1.3.14.) Evidently, if $\mathfrak{A}$ satisfies TB0 and is $\{s, t\}$-trivial, then $\mathfrak{A}$ satisfies TB1, ···, TB10. By 17(d), C1 gives a correspondence between right divisibility in $\mathcal{S}$ and right divisibility in the semigroup of operations $(s\alpha)$. Also, by 17(e), if $\mathfrak{A}$ satisfies TB0, ···, TB6 and if $\mathcal{S}$ is regular, then C1 is equivalent to the following condition, which can be expressed by a set of negations of equalities:

C′1. $\qquad (t\alpha)\mathbf{1} \nleqq (t\beta)\mathbf{1}$, provided $\beta \nleqq_R \alpha$.

Our final theorem (together with Lemma 7) shows that TB8, TB9, TB10 can often be replaced by conditions which are non-equational but conceptually simpler. Note that only TB6 and C1 among TB0, ···, TB6, C1, C2, C3 involve the structure of $\mathcal{S}$ (and that TB7 also does not involve the structure of $\mathcal{S}$).

THEOREM 32: *Assume that* $\mathcal{S}$ *has an identity and satisfies* S1, S2, S5, S9, *and that* $\mathfrak{A}$ *satisfies* TB0, $\cdots$, TB6 *and is not* $\{s, t\}$*-trivial. Then* $\mathfrak{A}$ *satisfies* TB8, TB9, *and* TB10 *if and only if it satisfies* C1, C2, and C3.

Proof: Assume the hypotheses of the theorem.[6] First assume that $\mathfrak{A}$ satisfies C1, C2, and C3. By C1 and TB0, the proviso of TB8 that γ is an $\leqq_R$ lub of $\{\alpha, \beta\}$ is equivalent to the following condition, which is internal to $\mathfrak{A}$: (i) $(t\gamma)\mathfrak{I}$ is the greatest element of $\{(t\delta)\mathfrak{I}: (t\delta)\mathfrak{I} \leqq (t\alpha)\mathfrak{I} \cdot (t\beta)\mathfrak{I}\}$.

Using TB0, 14(c), TB1, and TB0, one sees that $(t\alpha)(t\delta)x \leqq (t\alpha)(t\beta)x$ if and only if $(t\delta)x \leqq (s\alpha)(t\alpha)(t\beta)x$. Using C1, 16(a), and TB4, one then sees that the proviso of TB9 that β is an $\leqq_R$ lower bound of $\{\delta: \beta\alpha \leqq_R \delta\alpha\}$ is equivalent to the following condition, which is internal to $\mathfrak{A}$: (ii) $(t\beta)\mathfrak{I}$ is the greatest element of $\{(t\delta)\mathfrak{I}: (t\delta)\mathfrak{I} \leqq (s\alpha)(t\alpha)(t\beta)\mathfrak{I}\}$.

By C2, $(s\alpha)(t\beta)\mathfrak{I} = (t\delta')\mathfrak{I}$ for some δ'. Using 14(f) and 16(a) one sees that $(t\alpha)\mathfrak{I} \cdot (t\beta)\mathfrak{I} = (t\alpha)(s\alpha)(t\beta)\mathfrak{I} = (t\alpha)(t\delta')\mathfrak{I} = (t\delta'\alpha)\mathfrak{I}$. Hence $(t\alpha)\mathfrak{I} \cdot (t\beta)\mathfrak{I} = (t\delta)\mathfrak{I}$ for some δ. Evidently, $(t\alpha)\mathfrak{I} \cdot (t\beta)\mathfrak{I} \leqq (t\alpha)\mathfrak{I}$ and $(t\alpha)\mathfrak{I} \cdot (t\beta)\mathfrak{I} \leqq (t\beta)\mathfrak{I}$. Hence (i) implies that $(t\alpha)\mathfrak{I} \cdot (t\beta)\mathfrak{I} \leqq (t\gamma)\mathfrak{I}$. Therefore $\mathfrak{A}$ satisfies TB8.

By 16(a), $(t\alpha)(t\beta)\mathfrak{I} = (t\beta\alpha)\mathfrak{I}$. Hence, by C2, $(s\alpha)(t\alpha)(t\beta)\mathfrak{I} = (t\delta)\mathfrak{I}$ for some δ. Hence (ii) implies that $(t\beta)\mathfrak{I} = (s\alpha)(t\alpha)(t\beta)\mathfrak{I}$. Therefore $\mathfrak{A}$ satisfies TB9.

Assume that β is not an $\leqq_R$ lower bound of $\{\delta: \beta\alpha \leqq_R \delta\alpha\}$. Then, as we saw above, (ii) fails. By TB4 and 14(h), $-(s\alpha)(t\alpha) - (t\beta)\mathfrak{I} < (t\beta)\mathfrak{I}$. Now $x \cdot -(s\alpha)(t\alpha) - (t\beta)\mathfrak{I} \leqq -(s\alpha)(t\alpha) - (t\beta)\mathfrak{I}$. It follows from C3 that $(s\gamma)(x \cdot -(s\alpha)(t\alpha) - (t\beta)\mathfrak{I}) = x \cdot -(s\alpha)(t\alpha) - (t\beta)\mathfrak{I}$. Therefore $\mathfrak{A}$ satisfies TB10.

For the other half of the proof, assume now that $\mathfrak{A}$ satisfies TB8, TB9, and TB10. Since $\mathcal{S}$ satisfies S9, 19(a) implies that $\mathfrak{A}$ satisfies C2. Now suppose that $-(s\alpha)(t\alpha) - (t\beta)\mathfrak{I} < (t\beta)\mathfrak{I}$, and consider any $x \leqq -(s\alpha)(t\alpha) - (t\beta)\mathfrak{I}$, so that $x = x \cdot -(s\alpha)(t\alpha) - (t\beta)\mathfrak{I}$. By 14(h) and since $\mathfrak{A}$ satisfies TB9, β is not an $\leqq_R$ lower bound of $\{\delta: \beta\alpha \leqq_R \delta\alpha\}$. Since $\mathfrak{A}$ satisfies

---

[6] The conditions on $\mathcal{S}$ are not required for the first half of the proof.

TB10, therefore $(s\gamma)x = (s\gamma)(x \cdot -(s\alpha)(t\alpha) - (t\beta)\mathfrak{1}) = x \cdot -(s\alpha)(t\alpha) - (t\beta)\mathfrak{1} = x$. Therefore $\mathfrak{A}$ satisfies C3.

Finally, we use the analogue of Theorem 31 for the case where TB7 is not assumed to hold. According to this analogue, $\mathfrak{A}$ is isomorphic to an algebra which is induced by an antirepresentation $\hat{}$ of $\mathcal{S}$ such that either $\hat{}$ is trivial or else $\hat{}$ is $\langle \leqq_R, \leqq_L \rangle$-faithful and $\{\hat{\alpha}: \alpha \in \mathcal{S}\}$ has suitable ranges. Since $\mathfrak{A}$ is not $\{s, t\}$-trivial, it follows that $\hat{}$ is $\langle \leqq_R, \leqq_L \rangle$-faithful. By 6(b) and the regularity of $\mathcal{S}$ it follows that $\beta \leqq_R \alpha$ and $\operatorname{ran} \hat{\alpha} \subseteq \operatorname{ran} \hat{\beta}$ are equivalent. Since $\operatorname{ran} \hat{\alpha} = (t\alpha)\mathfrak{1}$ and $\operatorname{ran} \hat{\beta} = (t\beta)\mathfrak{1}$, therefore $\mathfrak{A}$ satisfies C1. ♦

Consider the following condition, suggested by C1 and 17(b).

C1$_L$. $\quad (s\alpha)(t\alpha) \leqq (s\beta)(t\beta)$ if and only if $\alpha \leqq_L \beta$.

In contrast to Theorem 32 one can show that there are many $\mathfrak{A}$ which are not $\{s, t\}$-trivial and satisfy TB0, $\cdots$, TB10 but do not satisfy C1$_L$. For such $\mathfrak{A}$ it follows from TB6 and 17(a) that the mapping of each $\alpha \in \mathcal{S}$ into $(s\alpha)$ is not one-to-one. For example, assume that $I = \{0, 1, 2\}$, $\mathcal{S} = {}^I I$, $\operatorname{ran} \alpha = \{0\}$, $\operatorname{ran} \beta = \{1\}$, $\hat{}$ is as in §1, and $\mathfrak{A}$ is that subalgebra of the complete algebra induced by $\hat{}$ which is generated by $\{(t\delta)\mathfrak{1}: \delta \in \mathcal{S}\}$. Then $\alpha \not\leqq_L \beta$. Also, one can verify that each element of A is a union of intersections of elements x such that either x or $-x$ is in $\{(t\delta)\mathfrak{1}: \delta \in \mathcal{S}\}$. One can then verify that $(s\alpha)(t\alpha)x = \mathfrak{1} = (s\beta)(t\beta)x$ for any $x \neq \varnothing$. Hence $(s\alpha)(t\alpha) = (s\beta)(t\beta)$. (Since $\alpha\check{}\alpha = \beta\check{}\beta$, $\alpha\alpha = \alpha$, and $\beta\beta = \beta$, it follows that $(s\alpha) = (s\alpha)(t\alpha)(t\alpha)(s\alpha) = (s\beta)(t\beta)(t\beta)(s\beta) = (s\beta)$ while $\alpha \neq \beta$.)

REFERENCES

1. Clifford, A. H. and G. B. Preston, *The Algebraic Theory of Semigroups*, vol. 1. Amer. Math. Soc., Providence, R.I., 1961, pp. 1–224.
2. Copeland, A. G., Sr., "Note on cylindric algebras and polyadic algebras." *Michigan Math. J.*, 3 (1955–56), 155–157.
3. Craig, W., *Logic in Algebraic Form. Three Languages and Theories*. Amsterdam: North-Holland, 1974.

3_D. Demaree, D. B., "Copeland algebras." *J. Symbolic Logic*, **37** (1972) 646–656.

4. Halmos, P. R., *Algebraic Logic.* New York: Chelsea, 1962.

5. Henkin, L., D. Monk, and A. Tarski, *Cylindric Algebras.* Part I. Amsterdam: North-Holland, 1971.

6. Howard, C. M., "An approach to algebraic logic." Doctoral dissertation, University of California, Berkeley, 1965.

7. Jónsson, B. and A. Tarski, "Boolean algebras with operators." Part I. *Amer. J. Math.*, **73** (1951), 891–939.

8. Jurie, P. F., "Notion de quasi-somme amalgamée; premières applications à l'algèbre booléienne polyadique." *Comptes Rendus de l'Académie des Sciences, Ser. A.*, **264** (1967), 1033–1036.

9. Lucas, Th., "Sur l'équivalence des algèbres cylindriques et polyadiques." *Bull. Soc. Math. Belg.*, **20** (1968), 236–263.

10. Monk, J. D., "On an algebra of sets of finite sequences." *J. Symbolic Logic*, **35** (1970), 19–28.

11. Quine, W. V., "Toward a calculus of concepts." *J. Symbolic Logic*, **1** (1936), 2–25.

CONNECTIONS BETWEEN COMBINATORIAL THEORY AND ALGEBRAIC LOGIC

J. Donald Monk

1. INTRODUCTION

Algebraic logic has arisen as a subdiscipline of algebra mirroring constructions and theorems of mathematical logic. It is similar in this respect to such fields as algebraic topology and algebraic geometry, where the main constructions and theorems are algebraic in nature but the main intuitions underlying them are respectively topological and geometric. The main intuitions underlying algebraic logic are, of course, those of formal logic. We shall describe in this first section the intuitive background of algebraic logic, and state some of the central definitions and results in this area. In later sections we give some constructions which arrive at the fundamental algebras of algebraic logic from entirely different sources, in fact from certain configurations which play a basic role in combinatorial theory. These connections between algebraic logic and combinatorial theory are rather unexpected (at least to this author). Their implications and deeper causes have not been fully explored.

It is our intent to give enough details in the constructions and theorems below so that a reader unversed in algebraic logic can follow the exposition and check those proofs which are presented. We only give proofs for results which are new, however.

Let $\mathfrak{L}$ be any first-order language. Thus $\mathfrak{L}$ has an infinite sequence of individual variables $v_0, v_1, \cdots$, logical constants—say $\neg, \rightarrow, \forall, =, -$ and non-logical constants—say a system $\langle R_i : i \in I \rangle$ of relation symbols, where R_i is of rank $\rho_i < \omega$ for each $i \in I$ (ω is the set of all non-negative integers). We assume as known the usual syntactic notions defined in terms of $\mathfrak{L}$, e.g., the notions of a formula, a sentence (formula without free occurrences of variables), the conjunction $\varphi \wedge \psi$ of formulas of $\mathfrak{L}$, the notion of a formal proof from a set of sentences, etc. Given a set Γ of sentences, we may call two formulas φ and ψ *equivalent under* Γ, in symbols $\varphi \equiv_\Gamma \psi$, provided that the biconditional $\varphi \leftrightarrow \psi$ is provable from Γ. The relation $\equiv_\Gamma$ is in fact an equivalence relation on the set of formulas. If we let A_Γ denote the set of all equivalence classes under Γ, we find that algebraic operations can be introduced on A_Γ which reflect the syntactic operations of building formulas:

$$
\begin{aligned}
[\varphi]_\Gamma + [\psi]_\Gamma &= [\varphi \vee \psi]_\Gamma, \\
[\varphi]_\Gamma \cdot [\psi]_\Gamma &= [\varphi \wedge \psi]_\Gamma, \\
-[\varphi]_\Gamma &= [\neg \varphi]_\Gamma, \\
c_i[\varphi]_\Gamma &= [\exists v_i \varphi]_\Gamma, \\
d_{ij} &= [v_i = v_j]_\Gamma.
\end{aligned}
$$

Here $[\varphi]_\Gamma$ is the equivalence class of φ under Γ. Note that d_{ij} is a 0-ary operation on A_Γ, i.e., an element of A_Γ. The algebra $\mathfrak{A}_\Gamma = \langle A_\Gamma, +, \cdot, -, c_i, d_{ij} \rangle_{i,j \in \omega}$ thus associated with $\mathfrak{L}$ and Γ is one of the fundamental algebras studied in algebraic logic. It turns out that most of the constructions and theorems of logic can be algebraically reflected using these algebras $\mathfrak{A}_\Gamma$. For example, Γ is complete and consistent iff $\mathfrak{A}_\Gamma$ is simple; the theorem that any consistent theory can be extended to a complete and consistent theory is mirrored by the theorem that any algebra $\mathfrak{A}_\Gamma$ with $|A_\Gamma| > 1$ has a simple homomorphic image.

The notion of a cylindric algebra is obtained from these algebras $\mathfrak{A}_\Gamma$ by a process of abstraction. Let α be any ordinal. A *cylindric algebra of dimension* α, for brevity a $\boldsymbol{CA}_\alpha$, is an algebraic structure $\mathfrak{A} = \langle A, +, \cdot, -, c_i, d_{ij}\rangle_{i,j<\alpha}$ satisfying the following conditions for all $i, j, k < \alpha$ and all $x, y \in A$:

(C_0) $\langle A, +, \cdot, - \rangle$ is a Boolean algebra,
(C_1) $c_i 0 = 0$,
(C_2) $x \leqq c_i x$,
(C_3) $c_i(x \cdot c_i y) = c_i x \cdot c_i y$,
(C_4) $c_i c_j x = c_j c_i x$,
(C_5) $d_{ii} = 1$,
(C_6) if $j \neq i, k$, then $c_j(d_{ij} \cdot d_{jk}) = d_{ik}$,
(C_7) if $i \neq j$, then $c_i(d_{ij} \cdot x) \cdot c_i(d_{ij} \cdot -x) = 0$.

The abstraction process is so familiar in modern mathematics that we do not have to describe its advantages. That this abstraction is sound is established by the following logical representation theorem, which is not very difficult to prove.

THEOREM 1.1: *For any algebra $\mathfrak{A}$ similar to $\boldsymbol{CA}_\omega$'s the following two conditions are equivalent:*

(i) $\mathfrak{A} \cong \mathfrak{A}_\Gamma$ *for some* Γ;
(ii) $\mathfrak{A}$ *is a* $\boldsymbol{CA}_\omega$ *such that* $|\{i: c_i x \neq x\}| < \omega$ *for all* $x \in A$.

Having at hand the abstract notion of a cylindric algebra, investigations in algebraic logic can proceed in two conceptually different, but actually closely related, ways. First, one can investigate the algebraic meaning of constructions and results of logic. Second, one can consider cylindric algebras as objects of investigation in their own right and discuss questions which naturally arise independently of any connection with logic. In this article we shall be concerned with problems of the second kind, specifically with certain methods of constructing $\boldsymbol{CA}$'s. These problems are, however, motivated by a central problem of the first kind, the set-theoretic representation problem. To describe this latter problem we need to discuss some basic concepts of model theory.

The most important concepts which are studied in mathematical logic concern the notion of a model. Let $\mathfrak{L}$ be a first-order language, as described above. A *model over* $\mathfrak{L}$ is a structure of the form $\mathfrak{A} = \langle A, R_i \rangle_{i \in I}$, where $A \neq 0$ and R_i is of rank ρ_i for each $i \in I$. We assume as known the notion of a sequence $x \in {}^\omega A$ *satisfying* a formula φ of $\mathfrak{L}$ in $\mathfrak{A}$, and such derivative notions as φ being *true* in $\mathfrak{A}$, $\mathfrak{A}$ being a *model* of a set Γ of sentences, etc. For each formula φ we set

$$\tilde{\varphi}^{(\mathfrak{A})} = \{x \in {}^\omega A : x \text{ satisfies } \varphi \text{ in } \mathfrak{A}\} = \tilde{\varphi} \text{ (if } \mathfrak{A} \text{ is understood)}.$$

Thus $\tilde{\varphi}^{(\mathfrak{A})}$ is a point-set in the ω-dimensional space ${}^\omega A$. Certain set-theoretic operations similar to the classical operations of descriptive set theory can be introduced corresponding to the basic syntactic operations:

$$\tilde{\varphi} \cup \tilde{\psi} = \widetilde{\varphi \vee \psi},$$

$$\tilde{\varphi} \cap \tilde{\psi} = \widetilde{\varphi \wedge \psi},$$

$${}^\omega A \sim \tilde{\varphi} = \widetilde{\neg \varphi},$$

$$C_i\tilde{\varphi} = \widetilde{\exists v_i \varphi} = \text{cylinder obtained by moving } \tilde{\varphi} \text{ parallel to the } i\text{-axis},$$

$$D_{ij} = \widetilde{v_i = v_j}.$$

The collection $\{\tilde{\varphi}^{\mathfrak{A}} : \varphi \text{ a formula of } \mathfrak{L}\}$ forms a $\boldsymbol{CA}_\omega$ under these operations. Again we make an abstraction from this notion to obtain a more general set-theoretic object. A *cylindric set algebra of dimension* α *with base* U, for short a $\boldsymbol{Cs}_\alpha^U$ or a $\boldsymbol{Cs}_\alpha$, is an algebraic structure

$$\mathfrak{A} = \langle A, \cup, \cap, \sim, C_i, D_{ij} \rangle_{i,j<\alpha}$$

such that A is a field of subsets of ${}^\alpha U$ closed under each C_i and with each D_{ij} as a member, where

$$D_{ij} = \{x \in {}^\alpha U : x_i = x_j\},$$

and for each $X \subseteq {}^\alpha U$,

$$C_i X = \{x \in {}^\alpha U : (\alpha \sim \{i\}) \upharpoonright x = (\alpha \sim \{i\}) \upharpoonright y \quad \text{for some} \quad y \in X\}.$$

In terms of this concept a purely algebraic form of the completeness theorem can be stated:

THEOREM 1.2: *If* $\mathfrak{A}$ *satisfies the condition* 1.1(ii), *then* $\mathfrak{A}$ *is homomorphic to a* $\boldsymbol{Cs}_\omega^U$ *for some* $U \neq 0$.

Proofs of 1.2 are somewhat deeper than those of 1.1.

The set-theoretic representation problem is the vaguely posed problem concerning possible improvements of 1.2. To make the problem more precise, let $\boldsymbol{R}_\alpha$ be the class of all $\boldsymbol{CA}_\alpha$'s isomorphic to a subdirect product of $\boldsymbol{Cs}_\alpha$'s; members of $\boldsymbol{R}_\alpha$ are called *representable* $\boldsymbol{CA}_\alpha$'s. In elementary terms the definition of $\boldsymbol{R}_\alpha$ runs as follows: $\mathfrak{A}$ is representable iff for every non-zero $x \in A$ there is a homomorphism h from $\mathfrak{A}$ onto a $\boldsymbol{Cs}_\alpha$ such that $hx \neq 0$. A very elementary argument shows that Theorem 1.2 is equivalent to the statement that every $\mathfrak{A}$ satisfying 1.1(ii) is an $\boldsymbol{R}_\alpha$. A more precise form of the representation problem is to describe properties of the class $\boldsymbol{R}_\alpha$, and give a useful characterization of it in abstract terms.

Some of the basic facts concerning $\boldsymbol{R}_\alpha$ are as follows. For $\alpha \leqq 1$, $\boldsymbol{CA}_\alpha = \boldsymbol{R}_\alpha$ and hence the representation problem is essentially solved. Next, $\boldsymbol{CA}_2 \neq \boldsymbol{R}_2$, but $\boldsymbol{R}_2$ can be characterized by $(C_0) - (C_7)$ together with the two equations

$$c_0[x \cdot y \cdot c_1(x \cdot -y)] \cdot -c_1(c_0 x \cdot -d_{01}) = 0,$$

$$c_1[x \cdot y \cdot c_0(x \cdot -y)] \cdot -c_0(c_1 x \cdot -d_{01}) = 0.$$

For each $\alpha > 2$, $\boldsymbol{CA}_\alpha \neq \boldsymbol{R}_\alpha$. The class $\boldsymbol{R}_\alpha$ is always a variety, but for $\alpha \geqq 3$ it is not finitely based, and for $\alpha \geqq \omega$ it cannot even be characterized by a certain natural kind of finite schema. For $\alpha \geqq \omega$, every simple $\boldsymbol{CA}_\alpha$ is representable. Theorem 1.2 gives a fundamental property of representable $\boldsymbol{CA}_\omega$'s. The last property of $\boldsymbol{R}_\alpha$'s which we will state will play a small role in section 4; its formulation requires a new concept. Let α and β be ordinals with $\alpha \leqq \beta$. Let

$$\mathfrak{A} = \langle A, +, \cdot, -, c_i, d_{ij} \rangle_{i,j<\alpha}$$

be a $\boldsymbol{CA}_\alpha$, and let

$$\mathfrak{B} = \langle B, +', \cdot', -', c_i', d_{ij}' \rangle_{i,j<\beta}$$

be a $\boldsymbol{CA}_\beta$. We say that $\mathfrak{A}$ *can be neatly embedded in* $\mathfrak{B}$ if there is an isomorphism f of $\mathfrak{A}$ into $\langle B, +', \cdot', -', c'_i, d'_{ij}\rangle_{i,j<\alpha}$ such that $c'_i fx = fx$ for all $x \in A$ and all $i \in \beta \sim \alpha$; f is called a *neat embedding of* $\mathfrak{A}$ *into* $\mathfrak{B}$.

THEOREM 1.3: *For any α the following two conditions are equivalent:*

(i) $\mathfrak{A} \in \boldsymbol{R}_\alpha$;
(ii) $\mathfrak{A}$ *can be neatly embedded in a* $\boldsymbol{CA}_{\alpha+\omega}$.

This completes our introduction to algebraic logic. A comprehensive treatment of the algebraic theory of cylindric algebras can be found in Henkin, Monk, Tarski [6]. The closely related theory of polyadic algebras is treated in Halmos [5].

2. QUASIGROUPS AND $\boldsymbol{CA}_3$'S

A *quasigroup* is an algebra $\mathfrak{A} = \langle A, \cdot\rangle$ such that for any $a, b \in A$ there is a unique x such that $x \cdot a = b$, and also a unique y such that $a \cdot y = b$. Quasigroups are essentially the same thing as latin squares; the latter form one of the main objects of study in combinatorial theory. For our present purposes, quasigroups are more convenient to deal with than latin squares. A good source of reference for quasigroups is Bruck [1]. Let $\mathfrak{A} = \langle A, \cdot\rangle$ be a quasigroup. We shall consider $\cdot$ as a certain ternary relation on A, in the usual way. If $X \subseteq \cdot$ and $i < 3$, we define

$$c_i X = \{y \in \cdot : y_i = x_i \quad \text{for some} \quad x \in X\}.$$

Further, let $q \in \cdot$. Then for $i, j < 3$ we let

$$d_{ij} = 1 \qquad \text{if} \qquad i = j,$$

$$d_{ij} = c_k\{q\} \qquad \text{if} \qquad \{i, j, k\} = 3.$$

By an $\mathfrak{A}, q - \boldsymbol{CA}_3$ we mean a system $\mathfrak{B} = \langle B, \cup, \cap, \sim, c_i, d_{ij}\rangle_{i,j<3}$ such that $\langle B, \cup, \cap, \sim\rangle$ is a Boolean algebra of subsets of $\cdot$, B is closed under c_i for each $i < 3$, and $d_{ij} \in B$ for all $i, j < 3$.

THEOREM 2.1: *If $\mathfrak{A} = \langle A, \cdot\rangle$ is a quasigroup and $q \in \cdot$, then any $\mathfrak{A}, q - \boldsymbol{CA}_3$ is a $\boldsymbol{CA}_3$.*

The proof of Theorem 2.1 is routine; let us verify (C_4) as an example. Suppose $\mathfrak{B}$ is an $\mathfrak{A}$, $q - \boldsymbol{CA}_3$, as above. Let $X \in B$ and assume $i, j < 3$. Obviously we may assume that $i \neq j$. If $X = 0$, clearly $c_i c_j X = 0 = c_j c_i X$. If, on the other hand, $X \neq 0$, we shall establish that $c_i c_j X = \cdot$ (and then by symmetry $c_j c_i X = \cdot = c_i c_j X$). So, let s be any member of $\cdot$. Choose $r \in X$. By the definition of quasigroup, there is then a unique $t \in \cdot$ such that $tj = rj$ and $ti = si$. Thus $t \in c_j X$, so $s \in c_i c_j X$, as desired. (We have actually established that $\mathfrak{B}$ is simple in the algebraic sense, but we do not need this fact below.)

Simple as it is, Theorem 2.1 turns out to have some uses in algebraic logic. Namely, we can use it to give an example of a non-representable $\boldsymbol{CA}_3$. To this end, consider the equation

(1) $$s_1^2 s_0^1 s_2^0 s_1^2 s_0^1 s_2^0 c_2 x = c_2 x,$$

where for brevity we let $s_j^i y = c_i(d_{ij} \cdot y)$ for all $i, j < 3$ and all y. The equation (1) is easily seen to hold in all cylindric set algebras, and hence also in all representable algebras. We shall now give an example of a $\boldsymbol{CA}_3$ in which it fails; thus this $\boldsymbol{CA}_3$ will be non-representable. Let $\mathfrak{A} = \langle A, \cdot \rangle$ be the quasigroup given by the following multiplication table:

$\cdot$	1	2	3	4	5	6
1	1	2	3	4	5	6
2	2	1	6	3	4	5
3	3	4	5	2	6	1
4	4	5	1	6	2	3
5	5	6	4	1	3	2
6	6	3	2	5	1	4

Let $q = \langle 2, 3, 6 \rangle$ and $X = \{\langle 5, 3, 4 \rangle\}$. Let $\mathfrak{B}$ be the $\mathfrak{A}, q - \boldsymbol{CA}_3$ of all subsets of $\cdot$. It is then easily verified that $\langle 2, 4, 3 \rangle \in s_1^2 s_0^1 s_2^0 s_1^2 s_0^1 s_2^0 c_2 X \sim c_2 X$. Thus, indeed, equation (1) fails to hold in $\mathfrak{B}$.

There are many natural questions that one can ask concerning our construction of a $\boldsymbol{CA}_3$ from a quasigroup. First, we can ask for a characterization of $\boldsymbol{CA}_3$'s isomorphic to an $\mathfrak{A}, q - \boldsymbol{CA}_3$. This question, although unresolved, is somewhat indefinite; a more definite form is as follows.

PROBLEM 1: *Let $\mathfrak{B}$ be a $\boldsymbol{CA}_3$ such that $c_0c_1x = c_0c_2x = c_1c_2x = 1$ for all non-zero $x \in B$. Is there a quasigroup $\mathfrak{A} = \langle A, \cdot \rangle$ and a $q \in \cdot$ such that $\mathfrak{B}$ is isomorphic to an $\mathfrak{A}, q - \boldsymbol{CA}_3$?*

Another natural question concerns relationships between algebraic properties of quasigroups and properties of associated $\boldsymbol{CA}_3$'s. The following result is of interest in this connection. A *loop* is a quasigroup $\mathfrak{A} = \langle A, \cdot \rangle$ having an identity element, i.e., having an element i such that $i \cdot a = a \cdot i = a$ for all $a \in A$. Thus the quasigroup given in the above table is actually a loop, but it is not a group.

THEOREM 2.2: *Let $\mathfrak{A} = \langle A, \cdot \rangle$ be a quasigroup.*

(i) *If $\mathfrak{A}$ is a group, $q \in \cdot$, and $\mathfrak{B}$ is any $\mathfrak{A}, q - \boldsymbol{CA}_3$, then $\mathfrak{B}$ is representable.*
(ii) *If $\mathfrak{A}$ is a loop but not a group, i is the identity of $\mathfrak{A}$, and $q = \langle i, i, i \rangle$, then the $\mathfrak{A}, q - \boldsymbol{CA}_3$ of all subsets of $\cdot$ is not representable.*

Proof: (i) Assume that $\mathfrak{A}$ is a group, let $q \in \cdot$, and let $\mathfrak{B}$ be an $\mathfrak{A}, q - \boldsymbol{CA}_3$. We shall assign to each $g \in \cdot$ a subset Fg of ${}^3A$:

$$Fg = \{x \in {}^3A : g_0 = q_0 x_1 x_2^{-1}, g_1 = x_2 x_0^{-1} q_1\}.$$

Clearly if $x \in Fg$, then g is uniquely determined by x. Thus

(1) $\quad g, h \in \cdot \quad$ and $\quad g \neq h \quad$ imply that $\quad Fg \cap Fh = 0.$

Also, obviously for any $x \in {}^3A$ there is a $g \in \cdot$ such that $x \in Fg$, so

(2) $$\bigcup_{g \in \cdot} Fg = {}^3A.$$

Finally,

(3) for any $g \in \cdot$ we have $Fg \neq 0$.

In fact, given $g \in \cdot$ we can let $x = \langle q_1 g_1^{-1}, q_0^{-1} g_0, i \rangle$, where i is the identity of $\mathfrak{A}$; clearly $x \in Fg$.

For any $b \in B$ we now let

$$Gb = \bigcup_{g \in b} Fg.$$

Using (1)–(3) it is easy to check that G is an isomorphism of the Boolean part of $\mathfrak{B}$ onto a field of subsets of ${}^3A$. It is straightforward to check that actually G is an isomorphism of $\mathfrak{B}$ itself onto a cylindric set algebra; for illustration we check that $Gc_0b \subseteq C_0Gb$ for any $b \in B$. Assume that $b \in B$ and $x \in Gc_0b$. Say $x \in Fg$ where $g \in c_0b$. Then choose $h \in b$ such that $g_0 = h_0$. Let $x_0' = q_1 h_1^{-1} x_2$, $y = \langle x_0', x_1, x_2 \rangle$. Now $h_0 = g_0 = q_0 x_1 x_2^{-1}$ since $x \in Fg$, and

$$x_2 x_0'^{-1} q_1 = x_2 x_2^{-1} h_1 q_1^{-1} q_1 = h_1.$$

It follows that $y \in Fh$. Since $h \in b$, thus $y \in Gb$, so $x \in C_0Gb$, as desired.

To prove (ii), suppose $\mathfrak{A}$ is a loop but is not a group. Thus $\cdot$ is not associative, so there exist elements $a, b, c \in A$ with $a \cdot (b \cdot c) \neq (a \cdot b) \cdot c$. Let i be the identity of $\mathfrak{A}$, set $q = \langle i, i, i \rangle$, and let $\mathfrak{B}$ be the $\mathfrak{A}, q - \boldsymbol{CA}_3$ of all subsets of $\cdot$. Now consider the following equation:

$$(4) \quad c_2\{c_0[c_2(c_0x \cdot c_1y) \cdot d_{02}] \cdot c_1z\}$$
$$= c_2[c_1(c_2\{c_0[c_2(c_1y \cdot d_{12}) \cdot d_{02}] \cdot c_1z\} \cdot d_{12}) \cdot c_0x].$$

It is easily verified that (4) holds in every cylindric set algebra and hence in every $\boldsymbol{R}_3$. However, (4) does not hold in $\mathfrak{B}$, and thus $\mathfrak{B}$ is not representable. In fact, if we let $x = \{\langle a, i, a \rangle\}$, $y = \{\langle i, b, b \rangle\}$, and $z = \{\langle i, c, c \rangle\}$, then it is easily verified that (4) fails.

COROLLARY 2.3: *If $\mathfrak{A}$ is a loop, i is the identity of $\mathfrak{A}$, $q = \langle i, i, i \rangle$, and $\mathfrak{B}$ is the $\mathfrak{A}, q - \boldsymbol{CA}_3$ of all subsets of $\cdot$, then a necessary and sufficient condition for $\mathfrak{A}$ to be a group is that $\mathfrak{B}$ is representable.*

Theorem 2.2 suggests the following variant of Problem 1:

PROBLEM 2: *Let $\mathfrak{B}$ be an $\boldsymbol{R}_3$ such that $c_0c_1x = c_0c_2x = c_1c_2x = 1$ for all non-zero $x \in B$. Is there a group $\mathfrak{A} = \langle A, \cdot \rangle$ and a $q \in \cdot$ such that $\mathfrak{B}$ is isomorphic to an $\mathfrak{A}, q - \boldsymbol{CA}_3$?*

It may be that methods of McKenzie [10] can be used to settle this question.

3. PROJECTIVE GEOMETRIES AND $\boldsymbol{CA}_3$'S

The construction we shall now describe was first carried out for relation algebras by Jónsson [8] and Lyndon [9]. Surprisingly, we do not know any connections between the present construction and that of section 2, although projective planes are essentially just complete systems of mutually orthogonal latin squares. By a projective geometry we understand a system $\langle \mathcal{P}, \mathcal{L} \rangle$ such that $\mathcal{P}$ is a non-empty set (of "points"), $\mathcal{L}$ is a non-empty collection of subsets (called "lines") of $\mathcal{P}$, and:

(G_1) each line contains at least four points;
(G_2) each pair of distinct points p and q lies on a unique line $\overline{pq}$;
(G_3) if p, q, r, and s are distinct points and $\overline{pq}$ and $\overline{rs}$ have a common point, then $\overline{pr}$ and $\overline{qs}$ have a common point (see Figure 1).

We shall assume a knowledge of elementary projective geometry; see, e.g., Seidenberg [16].

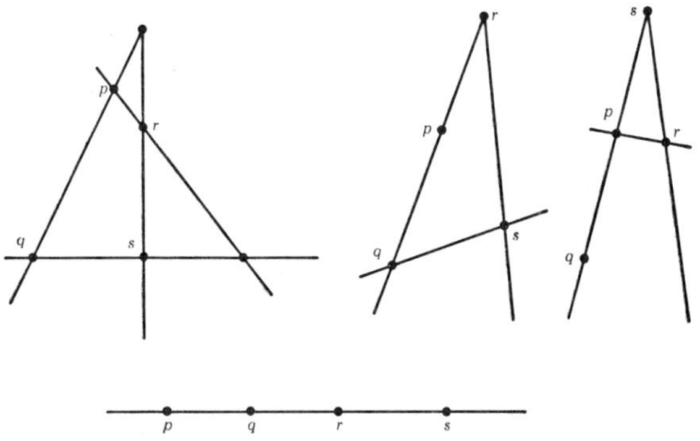

FIG. 1

If R is an equivalence relation on 3, let $R' = \{(i, j) : i, j < 3$ and $i \not R j\}$. (Here $3 = \{0, 1, 2\}$.) Let $\mathfrak{G} = \langle \mathcal{P}, \mathcal{L} \rangle$ be a projective geometry; we define $\mathfrak{U}\mathfrak{G}$ to be the collection of all pairs $\langle R, f \rangle$ satisfying the following conditions:

(1) R is an equivalence relation on 3;
(2) f maps R' into $\mathcal{P}$;
(3) for $(i, j) \in R', fij = fji$;
(4) if $i \not R j, i \not R k$, but $j R k$, then $fij = fik$;
(5) if $R =$ identity on 3, then either $f01 = f02 = f12$ or else $f01, f02, f12$ are distinct collinear points.

Now for $i, j < 3$ and $X \subseteq \mathfrak{U}\mathfrak{G}$ we let $3 = \{i, k, l\}$ and

$$c_iX = \{\langle R, f \rangle \in \mathfrak{U}\mathfrak{G} \text{: there is an } \langle S, g \rangle \in X \text{ such that } R \cap {}^2(3 \sim \{i\}) = S \cap {}^2(3 \sim \{i\}) \text{ and } fkl = gkl \text{ if } kR'l\};$$

$$d_{ij} = \{\langle R, f \rangle \in \mathfrak{U}\mathfrak{G} : i\, R\, j\}.$$

A $\mathfrak{G} - \boldsymbol{CA}_3$ is a structure $\langle A, \cup, \cap, \sim, c_i, d_{ij} \rangle_{i,j<3}$ such that A is a field of subsets of $\mathfrak{U}\mathfrak{G}$ closed under c_i for $i < 3$ and with $d_{ij} \in 3$ for $i, j < 3$. We denote by $\mathfrak{A}_\mathfrak{G}$ the $\mathfrak{G} - \boldsymbol{CA}_3$ of all subsets of $\mathfrak{U}\mathfrak{G}$.

A short description of the motivation behind this construction may be helpful. Intuitively speaking, we imagine $\mathfrak{G}$ embedded as a hyperplane in a space $\mathcal{H}$ of one higher dimension; $\langle R, f \rangle \in \mathfrak{U}\mathfrak{G}$, with $R =$ identity on 3, amounts to an abstract selection of three distinct points x_0, x_1, x_2 in $\mathcal{H} \sim \mathfrak{G}$ such that $\overline{x_ix_j}$ intersects $\mathfrak{G}$ at fij for $i, j < 3$, $i \neq j$. This accounts, for example, for the strange condition (5) since, by (G_3), if the lines $\overline{x_0x_1}$, $\overline{x_0x_2}$, $\overline{x_1x_2}$ are distinct, then they intersect $\mathfrak{G}$ in three distinct collinear points. The cylindrifications are restrictions of the natural set-theoretic cylindrifications. This intuitive description will become clearer during the proofs of the theorems below. For brevity we shall restrict ourselves mostly to the algebras $\mathfrak{A}_\mathfrak{G}$ instead of the more general $\mathfrak{G} - \boldsymbol{CA}_3$'s.

THEOREM 3.1: *If $\mathfrak{G}$ is a projective geometry, then $\mathfrak{A}_\mathfrak{G}$ is a* $\boldsymbol{CA}_3$.

We shall not give a detailed proof of 3.1, since it is straightforward and very close to the proof of Lemma 2.1 in Monk [12].

To illustrate the proof we shall establish (C_4). Let $\mathfrak{G} = \langle \mathcal{P}, \mathcal{L} \rangle$. Suppose $i, j < 3$ and $X \subseteq \mathfrak{U}\mathfrak{G}$. If $i = j$, obviously $c_i c_j X = c_j c_i X$. Hence assume $i \neq j$. Clearly $c_i c_j 0 = 0 = c_j c_i 0$. So, assume $X \neq 0$. Then, as in the proof of 2.1, we shall establish that $c_i c_j X = \mathfrak{U}\mathfrak{G}$; thus ($C_4$) follows by symmetry. Let, then, $\langle R, f \rangle \in \mathfrak{U}\mathfrak{G}$ be arbitrary. Choose $\langle S, g \rangle \in X$. Choose k so that $3 = \{i, j, k\}$. If $jR'k$ or $iS'k$ let

$$T = (R \cap {}^2\{j, k\}) \cup (S \cap {}^2\{i, k\}),$$

while if $j\, R\, k$ and $i\, S\, k$ let $T = {}^2 3$. Obviously in either case T is an equivalence relation on 3. If $T = {}^2 3$, let $h = 0$. In case $T \neq {}^2 3$, we have $i\, T'\, j$. If moreover $j\, T\, k$ and $i\, T'\, k$ we let $hij = hji = hik = hki = gik$; if $j\, T'\, k$ and $i\, T\, k$ let $hij = hji = hjk = hkj = fjk$. Finally, if $j\, T'\, k$ and $i\, T'\, k$, let $hjk = hkj = fjk$ and $hik = hki = gik$; further, if $fjk = gik$ let $hji = hij = fjk$, while if $fjk \neq gik$ let $hij = hji$ be a third point on the line $\overline{(fjk)(gik)}$. Clearly then $\langle T, h \rangle \in \mathfrak{U}\mathfrak{G}$, $\langle R, f \rangle \in c_i\{\langle T, h \rangle\}$, and $\langle T, h \rangle \in c_j\{\langle S, g \rangle\}$. Thus $\langle R, f \rangle \in c_i c_j X$, as desired. (Again we have actually established that $\mathfrak{A}_{\mathfrak{G}}$ is simple.)

Once more we shall be interested in the relationship between the representability of $\mathfrak{A}_{\mathfrak{G}}$ and properties of the geometry $\mathfrak{G}$.

THEOREM 3.2: *If $\mathfrak{G}$ is a hyperplane in a space $\mathfrak{G}'$ of one higher dimension, then $\mathfrak{A}_{\mathfrak{G}}$ is representable.*

We shall only outline the proof of 3.2, since it is similar to the proof of Lemma 2.2 in Monk [12]. Let $\mathfrak{G} = \langle \mathcal{P}, \mathcal{L} \rangle$ and $\mathfrak{G}' = \langle \mathcal{P}', \mathcal{L}' \rangle$. Set $U = \mathcal{P}' \sim \mathcal{P}$, and for any $x \in {}^3U$ set

$$R_x = \{(i, j) : i, j < 3, x_i = x_j\}.$$

Obviously R_x is an equivalence relation on 3. We define f_x mapping R'_x into $\mathcal{P}$ by setting, for any $(i, j) \in R'_x$, $f_x ij = \overline{x_i x_j} \cdot \mathcal{P}$. Finally, for any $X \in A_{\mathfrak{G}}$ we set

$$FX = \{x \in {}^3U : \langle R_x, f_x \rangle \in X\}.$$

It is now very straightforward to verify that F is an isomorphism from $\mathfrak{A}_{\mathfrak{G}}$ onto a cylindric set algebra of subsets of ${}^3U$. To illustrate, we shall check that F preserves c_i. Let $3 = \{i, j, k\}$.

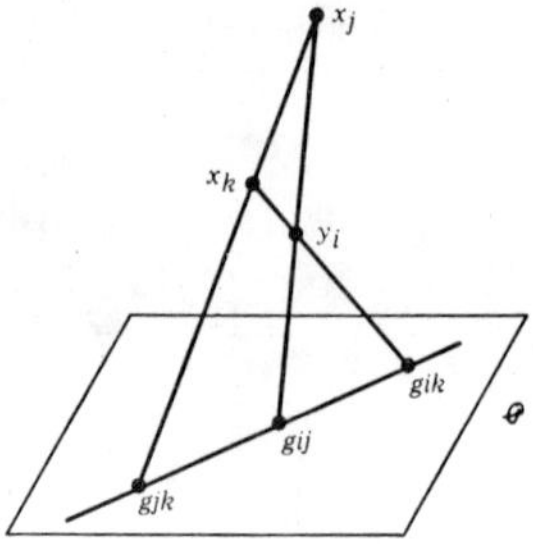

FIG. 2

First, suppose that $X \in A_{\mathfrak{G}}$ and $x \in Fc_iX$. Thus $\langle R_x, f_x \rangle \in c_iX$; choose, then, $\langle S, g \rangle \in X$ such that $R_x \cap {}^2(3 \sim \{i\}) = S \cap {}^2(3 \sim \{i\})$ and $f_xkj = gkj$ if kR'_xj. We will find $y \in {}^3U$ such that $(3 \sim \{i\}) \upharpoonright x = (3 \sim \{i\}) \upharpoonright y$ and $\langle R_y, f_y \rangle = \langle S, g \rangle$; this will prove the inclusion $Fc_iX \subseteq C_iFX$. The desired property of y is obvious from its definition in each of the following cases. Of course we let $y_j = x_j$, $y_k = x_k$.

Case 1. $S = {}^23$. Then $j\ R_x\ k$, so $x_j = x_k$. Let $y_i = x_j$.

Case 2. $j\ S\ k$, $i\ S'\ j$. Let y_i be a third point on the line $\overline{x_jg_{ij}}$.

Case 3. $j\ S'\ k$, $i\ S\ j$. (The case $j\ S'\ k$, $i\ S\ k$ is treated similarly.) Let $y_i = x_j$.

Case 4. $S =$ identity on 3, and $g01 = g02 = g12$. Let y_i be a point on the line $\overline{(gjk)x_j}$ different from gjk, x_j, x_k.

Case 5. $S =$ identity on 3, and $g01$, $g02$, $g12$ are distinct collinear points. Let $y_i = \overline{x_k(gik)} \cdot \overline{x_j(gij)}$ (see Figure 2).

Thus the inclusion $Fc_iX \subseteq C_iFX$ is established.

Now suppose that $x \in C_iFX$. Say $y \in FX$ and $(3 \sim \{i\}) \upharpoonright x = (3 \sim \{i\}) \upharpoonright y$. Thus $\langle R_y, f_y \rangle \in X$. Clearly also $\langle R_x, f_x \rangle \in c_i\{\langle R_y, f_y \rangle\} \subseteq c_iX$. Thus $x \in Fc_iX$, as desired.

Some special cases of Theorem 3.2 are worth mention. If $\mathfrak{G}$ has dimension 3 or more, then $\mathfrak{G}$ can always be embedded as a hyperplane in a space of one higher dimension, and hence $\mathfrak{A}_{\mathfrak{G}}$ is always representable. If $\mathfrak{G}$ has dimension 2, then $\mathfrak{G}$ can be so embedded iff $\mathfrak{G}$ is Desarguesian. Thus $\mathfrak{A}_{\mathfrak{G}}$ is representable whenever $\mathfrak{G}$ is a

Desarguesian projective plane. Finally, the seemingly trivial case of dimension one is of great importance. In this case, $\mathfrak{G}$ is merely a set with at least 4 points, and to say that $\mathfrak{G}$ can be embedded as a hyperplane in a space of one higher dimension is just to say that $\mathfrak{G}$ is a line in some projective plane. If $\mathfrak{G}$ is infinite, then this is always true, and hence by Theorem 3.2 $\mathfrak{A}_{\mathfrak{G}}$ is representable. If $\mathfrak{G}$ is finite, then the exact determination of when $\mathfrak{G}$ is a line in a projective plane is unresolved. If $\mathfrak{G}$ has $p^n + 1$ elements for some prime p and some $n > 0$, then $\mathfrak{G}$ is a line in a projective plane. But by a celebrated theorem of Bruck and Ryser (see [2]), there are infinitely many $\mathfrak{G}$ which cannot be a line in a projective plane.

We now wish to consider the converse of Theorem 3.2. We call a complete $\boldsymbol{CA}_\alpha$ $\mathfrak{A}$ *completely representable* if for every non-zero $a \in A$ there is a homomorphism h from $\mathfrak{A}$ into a $\boldsymbol{Cs}_\alpha$ such that $ha \neq 0$ and h carries arbitrary sums (joins) into unions. It is easily verified that a finite $\boldsymbol{CA}_\alpha$ is representable if and only if it is completely representable. Thus for finite $\mathfrak{G}$, the following theorem is an exact converse of Theorem 3.2.

THEOREM 3.3: *If $\mathfrak{A}_{\mathfrak{G}}$ is completely representable, then $\mathfrak{G}$ is a hyperplane in a space $\mathfrak{G}'$ of one higher dimension.*

Again, we shall not give a complete proof of Theorem 3.3; cf. the proof of Lemma 2.3 in [12], and the proof of Theorem 1 of [9]. We shall just define $\mathfrak{G}'$, and verify (G_3) for it under the assumption that (G_1) and (G_2) hold. Let H be a complete homomorphism from $\mathfrak{A}_{\mathfrak{G}}$ onto a $\boldsymbol{Cs}_3$ of subsets of ${}^3U$ such that $H\mathfrak{U}\mathfrak{G} \neq 0$. We may assume that $U \cap \mathcal{P} = 0$, where $\mathfrak{G} = \langle \mathcal{P}, \mathcal{L} \rangle$. Let $\mathfrak{G}' = \langle \mathcal{P}', \mathcal{L}' \rangle$, where $\mathcal{P}' = \mathcal{P} \cup U$, and $\mathcal{L}'$ consists of the lines in $\mathcal{L}$ together with all sets of the form

$$L(p, u) = \{p, u\} \cup \{v \in U : \langle v, u, u \rangle \in H\{\langle R, f \rangle\}\},$$

where $p \in \mathcal{P}$, $u \in U$, $R = \{(0,0), (1,1), (2,2), (1,2), (2,1)\}$, and $f01 = f10 = f02 = f20 = p$. First note

(1) $\qquad$ if $\quad 0 \neq X \in A_{\mathfrak{G}}, \quad$ then $\quad HX \neq 0.$

In fact, $c_0c_1X = \mathfrak{U}\mathfrak{G}$ by the proof of Theorem 3.1, so $Hc_0c_1X = C_0C_1HX = {}^3U$. Since $C_0C_10 = 0$, it follows that $HX \neq 0$.

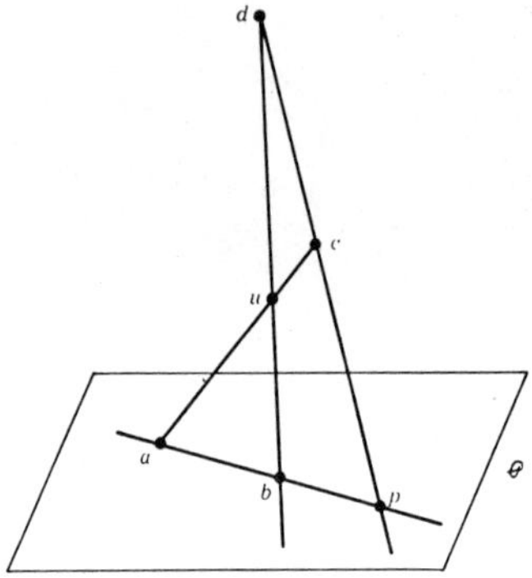

FIG. 3

Now we turn to the proof of (G_3), assuming that (G_1) and (G_2) hold. Let a, b, c, d be distinct points in $\mathcal{P}'$ such that $\overline{ab}$ and $\overline{cd}$ have a common point; we are to show that $\overline{ac}$ and $\overline{bd}$ have a common point. We may assume that $\overline{ab} \neq \overline{cd}$, and that $\overline{ab} \cdot \overline{cd}$ is different from a, b, c, d. If $\overline{ab}$ and $\overline{cd}$ are in $\mathcal{L}$, the desired conclusion follows since $\mathfrak{G}$ is a geometry.

Now suppose $\overline{ab}$ is in $\mathcal{L}$, while $\overline{cd}$ is not in $\mathcal{L}$ (the case $\overline{ab}$ not in $\mathcal{L}$, $\overline{cd}$ in $\mathcal{L}$ is similar). Clearly then $a, b \in \mathcal{P}$ while $c, d \in U$. Let $p = \overline{ab} \cdot \overline{cd}$. Thus $c \in L(p, d)$, so $\langle c, d, d\rangle \in H\{\langle R, f\rangle\}$, where $\langle R, f\rangle$ is as above. Now let S be the identity on 3, and let $g01 = g10 = p$, $g02 = g20 = a$, $g12 = g21 = b$. Note that $\{\langle R, f\rangle\} \subseteq c_2\{\langle S, g\rangle\}$; hence $H\{\langle R, f\rangle\} \subseteq C_2 H\{\langle S, g\rangle\}$. Thus, since $\langle c, d, d\rangle \in H\{\langle R, f\rangle\}$, it follows that there is a $u \in U$ such that $\langle c, d, u\rangle \in H\{\langle S, g\rangle\}$. (See Figure 3.) Next, let

$$h01 = h10 = h02 = h20 = a;$$

$$k01 = k10 = k02 = k20 = b.$$

Clearly

$$c_1\{\langle S, g\rangle\} \cap d_{12} = \{\langle R, h\rangle\},$$

$$c_2(c_0\{\langle S, g\rangle\} \cap d_{02}) \cap d_{12} = \{\langle R, k\rangle\};$$

since $\langle c, u, u\rangle \in C_1 H\{\langle S, g\rangle\} \cap D_{12}$ and

$$\langle u, d, d\rangle \in C_2(C_0 H\{\langle S, g\rangle\} \cap D_{02}) \cap D_{12},$$

it follows that $\langle c, u, u\rangle \in H\{\langle R, h\rangle\}$ and $\langle u, d, d\rangle \in H\{\langle R, k\rangle\}$. Hence $u \in L(a, c)$ and $d \in L(b, u)$. Therefore by (G_2) u is a common point of $\overline{ac}$ and $\overline{bd}$.

It remains to treat the case in which neither $\overline{ab}$ nor $\overline{cd}$ is in $\mathfrak{L}$. Here we will also consider several subcases. First suppose that $a \in \mathcal{P}$ and $c \in \mathcal{P}$ (the case $b \in \mathcal{P}$ and $d \in \mathcal{P}$ is similar); see Figure 4. Define

$$f01 = f02 = f10 = f20 = a;$$

$$g01 = g02 = g10 = g20 = c.$$

Let $\overline{ab} \cdot \overline{cd} = u$. Clearly $u \in U$, and $\langle u, b, b\rangle \in H\{\langle R, f\rangle\}$, $\langle u, d, d\rangle \in H\{\langle R, g\rangle\}$. Hence

$$\langle u, b, d\rangle \in C_2H\{\langle R, f\rangle\} \cap C_1H\{\langle R, g\rangle\}$$
$$= H(c_2\{\langle R, f\rangle\} \cap c_1\{\langle R, g\rangle\}).$$

Since H is completely additive, it follows that there is an $\langle S, h\rangle \in c_2\{\langle R, f\rangle\} \cap c_1\{\langle R, g\rangle\}$ such that $\langle u, b, d\rangle \in H\{\langle S, h\rangle\}$. Here S is the identity on 3, $h01 = a$, and $h02 = c$. Let $p = h12$. Then a, c, p are collinear since $\langle S, h\rangle \in \mathfrak{U}\mathfrak{S}$. Furthermore, if $k01 = k02 = k10 = k20 = p$, then

$$\{\langle R, k\rangle\} = c_1(c_0\{\langle S, h\rangle\} \cap d_{01}) \cap d_{12}.$$

Clearly $\langle b, d, d\rangle \in C_1(C_0H\{\langle S, h\rangle\} \cap D_{01}) \cap D_{12}$, so $\langle b, d, d\rangle \in H\{\langle R, k\rangle\}$. Thus $p = \overline{ac} \cdot \overline{bd}$, as desired.

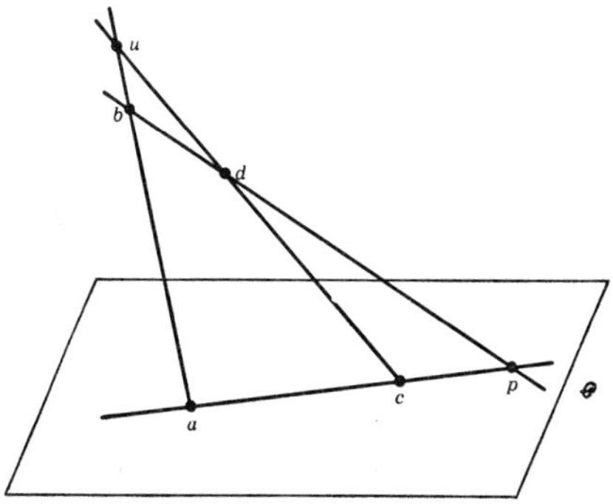

FIG. 4

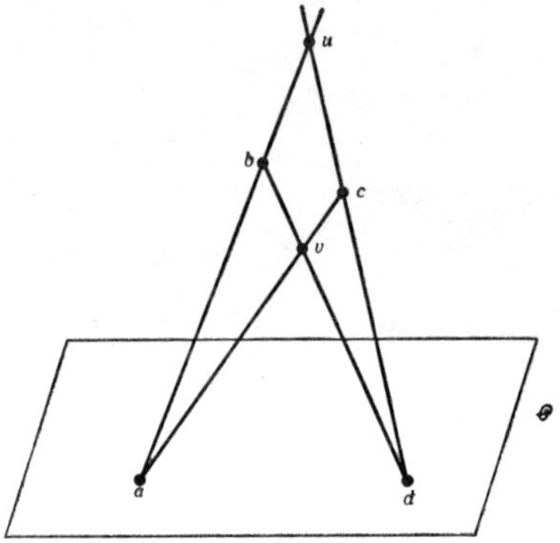

FIG. 5

Second, suppose that $a \in \mathcal{P}$ and $d \in \mathcal{P}$ (the case $b \in \mathcal{P}$ and $c \in \mathcal{P}$ is similar). (See Figure 5.) Let $u = \overline{ab} \cdot \overline{cd}$. Thus $u \in U$. Hence $\langle u, b, b\rangle \in H\{\langle R, f\rangle\}$ with $f01 = a$ and $\langle u, c, c\rangle \in H\{\langle R, g\rangle\}$ with $g01 = d$. Now let $x = c_0(c_2\{\langle R, f\rangle\} \cap c_1\{\langle R, g\rangle\})$. Then the following is easily established:

$$(1) \qquad d_{01} \cap c_1[d_{12} \cap c_2(d_{02} \cap x)] = x \cap d_{01}.$$

Now clearly $\langle b, b, c\rangle \in H(x \cap d_{01})$, so by (1), $\langle b, c, b\rangle \in Hx$. Hence there is a $v \in U$ such that $\langle v, c, b\rangle \in C_2H\{\langle R, f\rangle\}$ and $\langle v, c, b\rangle \in C_1H\{\langle R, g\rangle\}$. Obviously then v is a common point of $\overline{ac}$ and $\overline{bd}$, as desired.

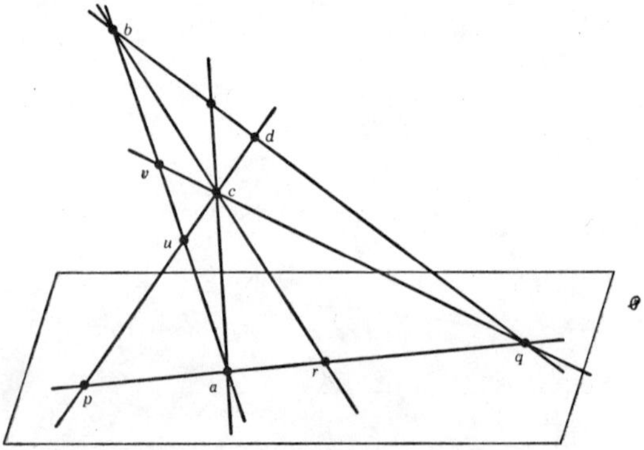

FIG. 6

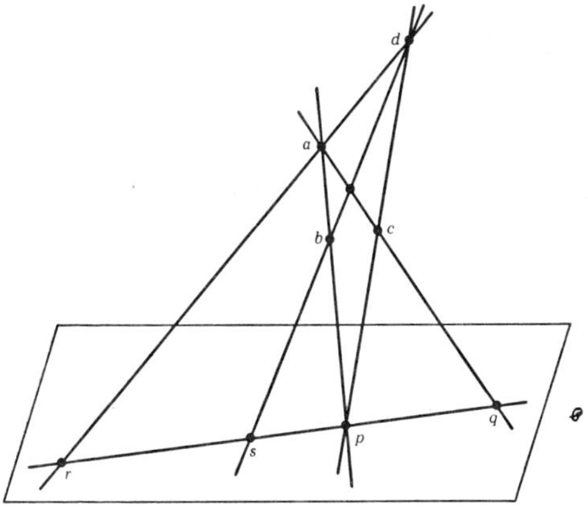

FIG. 7

We have now taken care of all cases in which two or more of the points a, b, c, d are in $\mathcal{P}$. Now suppose just one of them is in $\mathcal{P}$, say $a \in \mathcal{P}$. (See Figure 6.) Say $c, d \in L(p, c)$. Let $\overline{ab} \cdot \overline{cd} = u$. Then $\overline{pd} \cdot \overline{ab} = u$ also. Since both $a, p \in \mathcal{P}$, we apply a previous case to see that $\overline{pa}$ and $\overline{bd}$ intersect in some point q of $\mathcal{P}$. Thus $\overline{bq} \cdot \overline{cp} = d$; both $p, q \in \mathcal{P}$, so by a previous case $\overline{bc}$ and $\overline{pq}$ intersect in some point $r \in \mathcal{P}$. Thus $\overline{aq} \cdot \overline{bc} = r$, and $a, q \in \mathcal{P}$, so $\overline{ab} \cdot \overline{qc} = v$ for some v. Hence $\overline{ac} \cdot \overline{bq}$ exists; since $\overline{bq} = \overline{bd}$, the desired result follows.

Hence we may assume that none of a, b, c, d are in $\mathcal{P}$. Suppose $\overline{ab} \cdot \overline{cd} = p \in \mathcal{P}$. (See Figure 7.) Let $\overline{ac}$ intersect $\mathcal{P}$ at q. Then $\overline{aq} \cdot \overline{dp} = c$, so by a previous case $\overline{ad} \cdot \overline{pq} = r \in \mathcal{P}$ for some r. Also, $\overline{rd} \cdot \overline{pb} = a$, so $\overline{pr} \cdot \overline{bd} = s \in \mathcal{P}$ for some s. Finally, $\overline{da} \cdot \overline{sq} = r$, so $\overline{ds} \cdot \overline{aq} = \overline{bd} \cdot \overline{ac}$ exists, as desired. Finally, suppose $\overline{ab} \cdot \overline{cd} = u \notin \mathcal{P}$. (See Figure 8.) Let $\overline{ab} \cdot \mathcal{P} = p$, $\overline{cd} \cdot \mathcal{P} = q$. Now $\overline{pa} \cdot \overline{qc} = u$, so $\overline{pq} \cdot \overline{ac} = r \in \mathcal{P}$ for some r. Also, $\overline{pb} \cdot \overline{qd} = u$, so $\overline{pq} \cdot \overline{bd} = s \in \mathcal{P}$ for some s. Next, $\overline{pa} \cdot \overline{qd} = u$, so $\overline{pq} \cdot \overline{ad} = t \in \mathcal{P}$ for some t. $\overline{ad} \cdot \overline{rs} = t$, so $\overline{ar} \cdot \overline{ds} = \overline{ac} \cdot \overline{bd}$ exists, as desired. This completes the proof.

Theorems 3.2 and 3.3 can be used together to yield an important result concerning the class $\boldsymbol{R}_3$ of representable three-dimensional cylindric algebras. Namely, let $\boldsymbol{K}$ be an infinite collection of finite

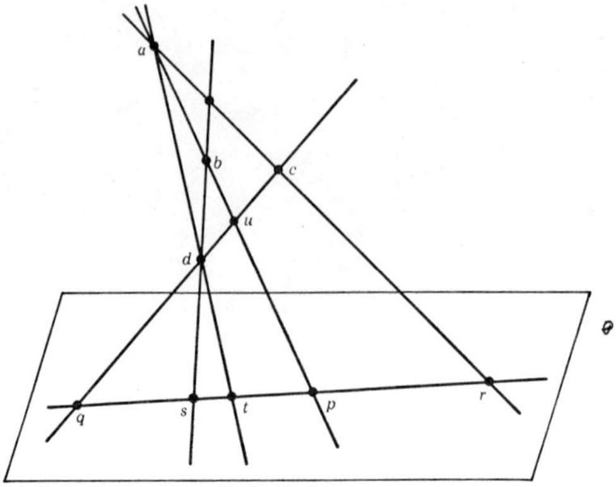

FIG. 8

one-dimensional projective geometries none of which can be embedded as a line in a plane; as mentioned above, such exist by [2]. Let F be a non-principal ultrafilter on $\boldsymbol{K}$. Then the following conditions hold:

(1) for each $\mathfrak{G} \in \boldsymbol{K}$, $\mathfrak{A}_{\mathfrak{G}}$ is non-representable (by Theorem 3.3);
(2) the ultraproduct $\mathfrak{B} = P_{\mathfrak{G}\in\boldsymbol{K}}\mathfrak{A}_{\mathfrak{G}}/F$ can be isomorphically embedded in $\mathfrak{A}_{\mathfrak{K}}$ for some infinite one-dimensional geometry $\mathfrak{K}$, and hence $\mathfrak{B}$ is representable.

From (1) and (2) it follows that $\boldsymbol{R}_3$ cannot be characterized by any finite set of first-order axioms. This is a special case of a more general result which can be established using results of the next section. For a proof of (2) see [12]; cf. also Monk [11] and McKenzie [10].

For the purpose of comparison with results in the next section we shall conclude this section with a purely combinatorial theorem of a simple nature. Let $S^2U = \{X : X \subseteq U, |X| = 2\}$.

THEOREM 3.4: *Let* $\mathfrak{G} = \langle \mathcal{P}, \mathcal{L} \rangle$ *be a one-dimensional projective geometry. Then the following conditions are equivalent:*

(i) $\mathfrak{G}$ *is a line in some projective plane;*
(ii) *there is a non-empty set* U *and a partition* $\langle T_p : p \in \mathcal{P} \rangle$ *of* S^2U *such that*

(a) *for all distinct* $u, v, w \in U$ *and all* $p \in \mathcal{P}$, *if* $\{u, v\} \in T_p$ *and* $\{v, w\} \in T_p$ *then* $\{u, w\} \in T_p$;

(b) *for all distinct* $u, v \in U$ *and all* $p, q, r \in \mathcal{P}$ *with* $|\{p, q, r\}| \neq 2$, *if* $\{u, v\} \in T_p$ *then there exists* $w \in U \sim \{u, v\}$ *such that* $\{u, w\} \in T_q$ *and* $\{w, v\} \in T_r$.

Proof: (i) $\Rightarrow$ (ii). Let $\mathfrak{G}$ be a line in a projective plane $\mathfrak{G}' = \langle \mathcal{P}', \mathcal{L}' \rangle$. Let $U = \mathcal{P}' \sim \mathcal{P}$, and for each $p \in \mathcal{P}$ let

$$T_p = \{\{u, v\} : u, v \in U, u \neq v, \quad \text{and} \quad \overline{uv} \cdot \mathcal{P} = p\}.$$

Clearly $\langle T_p : p \in \mathcal{P} \rangle$ is the desired partition of S^2U satisfying (a) and (b).

(ii) $\Rightarrow$ (i). Assume (ii). Let $\mathcal{P}' = \mathcal{P} \cup U$, where we assume that $\mathcal{P} \cap U = 0$. Let the lines of $\mathfrak{G}'$ be $\mathcal{P}$ together with all sets of the form

$$L(p, u) = \{v : \{u, v\} \in T_p\} \cup \{p, u\},$$

where $u \in U$ and $p \in \mathcal{P}$.

Since $|\mathcal{P}| \geqq 4$ by (G_1), using (b) we easily infer that

(1) any line of $\mathfrak{G}'$ has at least four points.

It is also clear from the definition of $L(p, u)$ that

(2) any two points of $\mathfrak{G}'$ lie on at least one line of $\mathfrak{G}'$.

(3) if $v \in V \cap L(p, u)$ with $u \in U$, then $L(p, u) = L(p, v)$.

To prove (3), first let $w \in L(p, u)$. We may assume that $u \neq v$. Thus, by hypothesis of (3), $\{u, v\} \in T_p$. If $w = u$, then obviously $w \in L(p, v)$. Also $w = v$ trivially yields $w \in L(p, v)$. Assume that $w \neq u$ and $w \neq v$. Then $\{u, w\} \in T_p$, so by (a) $\{v, w\} \in T_p$, and $w \in L(p, v)$ again. Therefore $L(p, u) \subseteq L(p, v)$. The converse is proved similarly.

Using (3) it is easy to check that

(4) any two distinct points of $\mathfrak{G}'$ lie on at most one line of $\mathfrak{G}'$.

To check that any two distinct lines intersect it suffices to take the distinct lines of the forms $L(p, u)$ and $L(q, v)$ with $p \neq q$ and

$u \neq v$. Choose $r \in \mathcal{P}$ with $\{u, v\} \in T_r$. If $r = p$, then $v \in L(p, u)$ and hence $L(p, u)$ and $L(q, v)$ intersect. Thus we may assume that $r \neq p$ and, similarly, that $r \neq q$. Then by (b) choose $w \in U \sim \{u, v\}$ so that $\{u, w\} \in T_p$ and $\{w, v\} \in T_q$. Thus $w \in L(p, u) \cap L(q, v)$, as desired. Thus

(5) any two distinct lines of $\mathfrak{G}'$ intersect.

From (1), (2), (4), (5) we see that $\mathfrak{G}'$ is a projective plane, as desired.

4. GRAPHS AND CA_α'S

The construction of section 3 can be slightly modified to yield a connection between certain graphs and CA_α's (now with $3 \leqq \alpha$). Throughout this section we assume that α is arbitrary but fixed, $3 \leqq \alpha$. If R is an equivalence relation on α, we let $R' = \{(i, j) : i, j < \alpha \text{ and } i \not{R}\, j\}$. A *selection set* for R is a subset T of α such that $i\, R'\, j$ whenever $i, j \in T$ and $i \neq j$. Now for any $\beta, \gamma < \omega$ we define $\mathfrak{C}(\alpha, \beta, \gamma)$ to be the collection of all pairs $\langle R, f \rangle$ satisfying the following conditions:

(1) R is an equivalence relation on α;
(2) f maps R' into γ;
(3) for $(i, j) \in R'$, $fij = fji$;
(4) if $i\, R'\, j$, $i\, R'\, k$, but $j\, R\, k$, then $fij = fik$;
(5) if T is a selection set for R and $|\, T \,| = \beta$, then

$$|\, \{ fij : i, j \in T, i \neq j\} \,| \neq 1.$$

If $\langle R, f \rangle \in \mathfrak{C}(\alpha, \beta, \gamma)$, then f can be considered as an edge coloring of the complete graph on α/R vertices using γ colors, such that no subcomplete-graph on β vertices is monochromatic (edges all of the same color). This will play an implicit role later. Now we construct a CA_α from $\mathfrak{C}(\alpha, \beta, \gamma)$ analogously to section 3. For

$i, j < \alpha$ and $X \subseteq \mathfrak{C}(\alpha, \beta, \gamma)$ we set

$$c_i X = \{\langle R, f\rangle \in \mathfrak{C}(\alpha, \beta, \gamma): \text{there is an } \langle S, g\rangle \in X \text{ such that } R \cap {}^2(\alpha \sim \{i\}) = S \cap {}^2(3 \sim \{i\}) \text{ and } fkl = gkl \text{ whenever } k\, R'\, l \text{ and } k, l \neq i\};$$

$$d_{ij} = \{\langle R, f\rangle \in \mathfrak{C}(\alpha, \beta, \gamma): i\, R\, j\}.$$

Finally, we let $A_{\alpha\beta\gamma}$ be the collection of all subsets of $\mathfrak{C}(\alpha, \beta, \gamma)$ and

$$\mathfrak{A}_{\alpha\beta\gamma} = \langle A_{\alpha\beta\gamma}, \cup, \cap, \sim, c_i, d_{ij}\rangle_{i,j<\alpha}.$$

The algebras $\mathfrak{A}_{\alpha 3\gamma}$ have been discussed at some length in Monk [13], some of the results of which will be generalized here. (Different generalizations have appeared in Demaree [3], Johnson [7] and Monk [14].) First of all, in fact, the following two results have identical proofs with those of Theorem 1.1 and 1.2 of [13]:

THEOREM 4.1: *If* $\gamma \geqq \alpha - 1$ *and* $\beta \geqq 3$, *then* $\mathfrak{A}_{\alpha\beta\gamma}$ *is a* $\mathbf{CA}_\alpha$.

THEOREM 4.2: *If* $3 \leqq \alpha \leqq \delta$, $\beta \geqq 3$, $\gamma \geqq \delta - 1$, *then* $\mathfrak{A}_{\alpha\beta\gamma}$ *is neatly embeddable in* $\mathfrak{A}_{\delta\beta\gamma}$.

By virtue of Theorem 1.3, we may interpret the conclusion of Theorem 4.2 as saying that $\mathfrak{A}_{\alpha\beta\gamma}$ is "approximately" representable. If $\delta = \alpha + \omega$, then $\mathfrak{A}_{\alpha\beta\gamma}$ really is representable. We shall need the following supplement to 4.1; general algebraically it expresses the fact that $\mathfrak{A}_{\alpha\beta\gamma}$ is simple (see [6]).

THEOREM 4.3: *If* $\alpha < \omega$, $\gamma \geqq \alpha - 1$, $\beta \geqq 3$, *and* $0 \neq X \subseteq \mathfrak{C}(\alpha, \beta, \gamma)$, *then* $c_0 \cdots c_{\alpha-1} X = \mathfrak{C}(\alpha, \beta, \gamma)$.

Proof: It obviously suffices to prove the following statement:

(6) if $\langle R, f\rangle, \langle S, g\rangle \in \mathfrak{C}(\alpha, \beta, \gamma)$ then $\langle R, f\rangle \in c_0 \cdots c_{\alpha-1}\{\langle S, g\rangle\}$.

To prove (6) we define a sequence $\langle\langle T_i, h_i\rangle: i \leqq \alpha\rangle$ of elements of $\mathfrak{C}(\alpha, \beta, \gamma)$ in such a way that the following conditions hold for each $i \leqq \alpha$:

(7) $T_i \cap {}^2(\alpha \sim i) = R \cap {}^2(\alpha \sim i)$ and $T_i \cap {}^2i = S \cap {}^2i$;
(8) if $j, k \in \alpha \sim i$ and $j\ R'\ k$, then $h_i jk = fjk$;
(9) if $j, k \in i$ and $j\ S'\ k$, then $h_i jk = gjk$;
(10) if $i > 0$, then $\langle T_{i-1}, h_{i-1}\rangle \in c_{i-1}\{\langle T_i, h_i\rangle\}$.

First we set $\langle T_0, h_0\rangle = \langle R, f\rangle$; obviously (7)–(10) hold then with $i = 0$. Now suppose that $\langle T_i, h_i\rangle$ has been defined so that (7)–(10) hold. We set

$$T_{i+1} = [T_i \cap^2 (\alpha \sim \{i\})] \cup \{(i,i)\} \cup \{(i,k), (k,i) : k < i, i\ S\ k\}$$
$$\cup \{(i,j), (j,i) : j > i \text{ and there is a } k < i \text{ with } j\ T_i\ k\ S\ i\}.$$

Obviously T_{i+1} is symmetric and reflexive on α. To show that it is transitive it is enough, by symmetry, to consider the following cases:

Case 1. $j\ T_{i+1}\ k\ T_{i+1}\ i$ with $j, k < i$. Then $j\ T_i\ k\ S\ i$ and hence $j\ S\ k\ S\ i$ by (7). Thus $j\ S\ i$, so $j\ T_{i+1}\ i$.

Case 2. $j\ T_{i+1}\ k\ T_{i+1}\ i$ with $j < i$, $k > i$. Then there is an $l < i$ with $j\ T_i\ k\ T_i\ l\ S\ i$; thus $j\ T_i\ l\ S\ i$, so $j\ T_{i+1}\ i$ as in Case 1.

Case 3. $j\ T_{i+1}\ k\ T_{i+1}\ i$ with $j > i$, $k < i$. Then $j\ T_i\ k\ S\ i$, so obviously $j\ T_{i+1}\ i$.

Case 4. $j\ T_{i+1}\ k\ T_{i+1}\ i$ with $j, k > i$. This case is similar to Case 2.

Case 5. $j\ T_{i+1}\ i\ T_{i+1}\ k$ with $j, k < i$. Then $j\ S\ i\ S\ k$, so $j\ S\ k$ and hence $j\ T_i\ k$ by (7), and $j\ T_{i+1}\ k$.

Case 6. $j\ T_{i+1}\ i\ T_{i+1}\ k$ with $j < i$, $k > i$. Then there is an $l < i$ with $j\ S\ i$, $k\ T_i\ l\ Si$. Thus $l\ S\ j$ so $l\ T_i\ j$ by (7), hence $k\ T_i\ j$ and hence $j\ T_{i+1}\ k$.

Case 7. $j\ T_{i+1}\ i\ T_{i+1}\ k$ with $j, k > i$. Then there exist $l, m < i$ with $j\ T_i\ l\ S\ i$ and $k\ T_i\ m\ S\ i$. Thus $l\ S\ m$, so $l\ T_i\ m$ by (7). Hence $j\ T_i\ k$, so $j\ T_{i+1}\ k$.

This establishes that T_{i+1} is transitive, and hence is an equivalence relation on α. Further, (7) for $i + 1$ is now obvious. Now if $j, k \in \alpha \sim \{i\}$ and $j\, T'_i\, k$ (hence $j\, T'_{i+1}\, k$), we set $h_{i+1}jk = h_i jk$. If $j < i$ and $j\, T'_{i+1}\, i$ (hence $j\, S'\, i$) we set $h_{i+1}ij = h_{i+1}ji = gji$. Finally, if $j > i$ and $j\, T'_{i+1}\, i$, we consider several cases.

Case 1). There is a $k \neq i$ such that $k\, T_{i+1}\, i$. Then we set $h_{i+1}ij = h_{i+1}ji = h_i jk$; clearly this does not depend on our particular choice of such a k.

Case 2). There is no $k \neq i$ such that $k\, T_{i+1}\, i$, but there is an $l < i$ with $l\, T_{i+1}\, j$. Then we set $h_{i+1}ij = h_{i+1}ji = gil$; again it is easy to check that this does not depend on our particular choice of such an l.

Case 3). There is no $k \neq i$ such that $k\, T_{i+1}\, i$, and there is no $l < i$ with $l\, T_{i+1}\, j$. Let $j_1, \cdots, j_m$ be a sequence of members of $\alpha \sim (i + 1)$ satisfying the following conditions:

$j_s\, T'_{i+1}\, j_t$ for $1 \leqq s < t \leqq m$;

for each s with $1 \leqq s \leqq m$, there is no $l < i$ such that $l\, T_{i+1}\, j_s$; if $u \in \alpha \sim (i + 1)$ and there is no $l < i$ such that $l\, T_{i+1}\, u$, then there is an s with $1 \leqq s \leqq m$ such that $u\, T_{i+1}\, j_s$.

Since $m \leqq \alpha - i - 1$ and $\gamma \geqq \alpha - 1$, we may pick distinct elements

$$h_{i+1}ij_1, \cdots, h_{i+1}ij_m \in \gamma \sim \{h_{i+1}it : t < i, i\, T'_{i+1}\, t\}.$$

Then we set $h_{i+1}j_si = h_{i+1}ij_s$ for all s with $1 \leqq s \leqq m$. Furthermore, if $u \in \alpha \sim (i + 1)$, there is no $l < i$ such that $l\, T_{i+1}\, u$, and $u \notin \{j_s : 1 \leqq s \leqq m\}$, then there is a unique s with $1 \leqq s \leqq m$ such that $u\, T_{i+1}\, j_s$, and we set

$$h_{i+1}ui = h_{i+1}iu = h_{i+1}ij_s.$$

In this way each element j as in Case 3 is taken care of. Thus h_{i+1} is now completely defined.

Now if $j, k \in \alpha \sim (i + 1)$, then $j, k \in \alpha \sim i$ and hence

$$h_{i+1}jk = h_i jk = fjk \text{ by (8) for } i.$$

Thus (8) holds for $i + 1$. Next, if $j, k \in i$ and $j\ S'\ k$, then

$$h_{i+1}\,jk = h_i\,jk = gjk;$$

to check that (9) holds we thus need only look at an element $h_{i+1}ij$, where $j < i$ and $i\ S'\ j$. Thus $i\ T'_{i+1}\ j$, so $h_{i+1}ij = gij$, as desired.

We still have to check that $\langle T_{i+1}, h_{i+1}\rangle \in \mathfrak{C}(\alpha, \beta, \gamma)$; it will then be obvious that (10) holds. Clearly here we only need to concern ourselves with conditions (4) and (5). To check condition (4), several cases need to be considered.

Case (1). $i\ T'_{i+1}\ j,\ \ i\ T'_{i+1}\ k,\ \ j\ T_{i+1}\ k,\ \ j, k < i$. This case is obvious since $\langle S, g\rangle \in \mathfrak{C}(\alpha, \beta, \gamma)$.

Case (2). $i\ T'_{i+1}\ j, i\ T'_{i+1}\ k, j\ T_{i+1}\ k, j < i, k > i$, and there is an $l < i$ such that $l\ T_{i+1}\ i$. Then $l\ S\ i$ and

$$\begin{aligned} h_{i+1}ik &= h_ilk && \text{by Case 1)}\\ &= h_ilj && \text{since } \langle T_i, h_i\rangle \in \mathfrak{C}(\alpha, \beta, \gamma)\\ &= glj && \text{by (9)}\\ &= gij && \text{since } \langle S, g\rangle \in \mathfrak{C}(\alpha, \beta, \gamma)\\ &= h_{i+1}ij. \end{aligned}$$

Case (3). $i\ T'_{i+1}\ j, i\ T'_{i+1}\ k, j\ T_{i+1}\ k, j < i, k > i$, and there is an $l > i$ such that $l\ T_{i+1}\ i$. Then by the definition of T_{i+1} there is a $u < i$ with $l\ T_i\ u\ S\ i$. Thus $u\ T_{i+1}\ i$, and the proof runs as in Case (2).

Case (4). $i\ T'_{i+1}\ j,\ i\ T'_{i+1}\ k,\ j\ T_{i+1}\ k,\ j < i,\ k > i$, and there is no $l \neq i$ such that $l\ T_{i+1}\ i$. Then $h_{i+1}ik = gij = h_{i+1}ij$ by Case 2).

Case (5). $i\ T'_{i+1}\ j, i\ T'_{i+1}\ k, j\ T_{i+1}\ k, j > i, k > i$, and there is an $l < i$ such that $l\ T_{i+1}\ i$. Then $l\ S\ i$ and

$$\begin{aligned} h_{i+1}ik &= h_ilk && \text{by Case 1)}\\ &= h_ilj && \text{since } \langle T_i, h_i\rangle \in \mathfrak{C}(\alpha, \beta, \gamma)\\ &= h_{i+1}ij && \text{by Case 1)} \end{aligned}$$

Case (6). $i\,T'_{i+1}\,j,\ i\,T'_{i+1}\,k,\ j\,T_{i+1}\,k,\ j > i,\ k > i$, and there is an $l > i$ such that $l\,T_{i+1}\,i$. This is treated similarly to Case (3).

Case (7). $j\,T'_{i+1}\,i,\ j\,T'_{i+1}\,k,\ i\,T_{i+1}\,k,\ j, k < i$. Then $i\,S\,k$ and

$$h_{i+1}\,ji = gji = gjk = h_i\,jk = h_{i+1}\,jk.$$

Case (8). $j\,T'_{i+1}\,i,\ j\,T'_{i+1}\,k,\ i\,T_{i+1}\,k,\ j < i,\ k > i$. Say $u < i$ and $k\,T_i\,u\,S\,i$. Then

$$h_{i+1}\,ji = gji = gju = h_i\,ju = h_i\,jk = h_{i+1}\,jk.$$

Case (9). $j\,T'_{i+1}\,i,\ j\,T'_{i+1}\,k,\ i\,T_{i+1}\,k,\ j > i,\ k \neq i$. Then $i\,S\,k$ and $h_{i+1}\,ji = h_i\,jk = h_{i+1}\,jk$.

Thus (4) has been established for $\langle T_{i+1}, h_{i+1}\rangle$. Turning to (5), let U be a selection set for T_{i+1} with $|\,T\,| = \beta$. If $i \notin U$, then the desired conclusion follows since $\langle T_i, h_i\rangle \in \mathfrak{C}(\alpha, \beta, \gamma)$. If $i \in U$ and there is a $k \neq i$ such that $k\,T_{i+1}\,i$, the desired conclusion again easily follows. Hence suppose $i \in U$ and there is no $k \neq i$ such that $k\,T_{i+1}\,i$. If for each $j \in U \sim \{i\}$ there is a $k < i$ with $k\,T_{i+1}\,j$, then the desired conclusion follows since $\langle S, g\rangle \in \mathfrak{C}(\alpha, \beta, \gamma)$. Finally, Case 3) takes care of the remaining possibility.

Thus $\langle T_{i+1}, h_{i+1}\rangle \in \mathfrak{C}(\alpha, \beta, \gamma)$, and our construction of $\langle\langle T_i, h_i\rangle : i < \alpha\rangle$ is complete. By (10), $\langle R, f\rangle \in c_0 \cdots c_{\alpha-1}\langle T_\alpha, h_\alpha\rangle$, and by (7) and (9), $\langle T_\alpha, h_\alpha\rangle = \langle S, g\rangle$, so the theorem is proved.

COROLLARY 4.4: *If* $\alpha < \omega$, $\gamma \geqq \alpha - 1$, $\beta \geqq 3$, *and* $\mathfrak{A}_{\alpha\beta\gamma}$ *is representable, then* $\mathfrak{A}_{\alpha\beta\gamma}$ *is isomorphic to a cylindric set algebra.*

Proof: If H is a homomorphism from $\mathfrak{A}_{\alpha\beta\gamma}$ onto a $\boldsymbol{Cs}_\alpha^U$ with $U \neq 0$ (and such an H must exist, by assumption), then $H\mathfrak{C}(\alpha, \beta, \gamma) = {}^\alpha U$. Hence by 4.3 $HX \neq 0$ whenever $X \neq 0$. Thus H is an isomorphism.

We now want to give some analogs of Theorem 3.4 for our algebras $\mathfrak{A}_{\alpha\beta\gamma}$; as will be seen, these theorems give connections between combinatorial questions and representation problems which are analogous to the connections given in section 3. Before giving our general result in this direction we first give a special result whose formulation is much simpler. The special result is taken from the unpublished work Demaree [3], and is included with his permission. Its proof is generalized for 4.6, and should serve to illustrate the more complicated proof.

THEOREM 4.5: *Let $\gamma \geqq 3$. Then the following two conditions are equivalent:*

(i) $\mathfrak{A}_{33\gamma}$ *is representable;*
(ii) *there is a non-empty set U and a partition $\langle T_\delta : \delta < \gamma \rangle$ of S^2U such that*

(a) *no T_δ contains a triangle, i.e., for all distinct $u, v, w \in U$ and all $\delta < \gamma$, if $\{u, v\} \in T_\delta$ and $\{u, w\} \in T_\delta$ then $\{u, w\} \notin T_\delta$;*
(b) *for all distinct $u, v \in U$ and all $\delta, \epsilon, \zeta < \gamma$ with $|\{\delta, \epsilon, \zeta\}| \geqq 2$, if $\{u, v\} \in T_\delta$ then there exists a $w \in U \sim \{u, v\}$ such that $\{u, w\} \in T_\epsilon$ and $\{w, v\} \in T_\zeta$.*

Proof: (i) $\Rightarrow$ (ii). By Corollary 4.4, let F be an isomorphism from $\mathfrak{A}_{33\gamma}$ onto a $\boldsymbol{Cs}^U$, where $U \neq 0$. Let P and R be the equivalence relations on 3 associated respectively with the partitions $\{\{0\}, \{1\}, \{2\}\}$ and $\{\{0\}, \{1, 2\}\}$; and for each $\delta < \gamma$ let f_δ be the mapping of R' into γ such that $f_\delta 01 = \delta$ and (3), (4) hold. Then

$$(11) \quad \text{if} \quad \langle u, v, w \rangle \in F\{\langle R, f_\delta \rangle\}, \quad \text{then} \quad v = w \quad \text{and} \quad \langle v, u, u \rangle \in F\{\langle R, f_\delta \rangle\}.$$

In fact, since $\langle R, f_\delta \rangle \in d_{12}$ it is clear that $v = w$. Also, it is easily checked that

$$\begin{aligned} \langle v, u, u \rangle &\in D_{12} \cap C_2[D_{02} \cap C_0(D_{01} \cap {\sim} D_{12} \cap C_1 F\{\langle R, f_\delta \rangle\})] \\ &= F(d_{12} \cap c_2[d_{02} \cap c_0(d_{01} \cap {\sim} d_{12} \cap c_1\{\langle R, f_\delta \rangle\})]) \\ &= F\{\langle R, f_\delta \rangle\}, \end{aligned}$$

as desired in (11). Now for each $\delta < \gamma$ let

$$T_\delta = \{\{u, v\} \in S^2U : \langle u, v, v \rangle \in F\{\langle R, f_\delta \rangle\}\}.$$

Clearly, then, $\langle T_\delta : \delta < \gamma \rangle$ is a partition of S^2U. To check (a) suppose u, v, and w are distinct members of U with $\{u, v\}$, $\{v, w\}$,

$\{u, w\} \in T_\delta$. Thus, by (11),

$$\langle u, v, v\rangle, \langle v, w, w\rangle, \langle u, w, w\rangle \in F\{\langle R, f_\delta\rangle\}.$$

Hence, as is easily seen,

$$\begin{aligned}\langle u, v, w\rangle &\in C_2F\{\langle R, f_\delta\rangle\} \cap C_1F\{\langle R, f_\delta\rangle\} \cap C_0(D_{01} \cap C_1F\{\langle R, f_\delta\rangle\}) \\ &= F[c_2\{\langle R, f_\delta\rangle\} \cap c_1\{\langle R, f_\delta\rangle\} \cap c_0(d_{01} \cap c_1\{\langle R, f_\delta\rangle\})];\end{aligned}$$

on the other hand, it is easily checked that

$$c_2\{\langle R, f_\delta\rangle\} \cap c_1\{\langle R, f_\delta\rangle\} \cap c_0(d_{01} \cap c_1\{\langle R, f_\delta\rangle\}) = 0.$$

This contradiction shows that (a) must hold.

To check (b), suppose $u \neq v$, $\delta, \epsilon, \zeta < \gamma$, $|\{\delta, \epsilon, \zeta\}| \geqq 2$, and $\{u, v\} \in T_\delta$. Thus $\langle u, v, v\rangle \in F\{\langle R, f_\delta\rangle\}$. Now let g map P' into γ in such a way that $g01 = \delta$, $g02 = \epsilon$, and $g12 = \zeta$. Thus $\langle P, g\rangle \in \mathfrak{C}(3, 3, \gamma)$. Clearly $\langle R, f_\delta\rangle \in c_2\{\langle P, g\rangle\}$, so

$$\langle u, v, v\rangle \in F\{\langle R, f_\delta\rangle\} \subseteq C_2F\{\langle P, g\rangle\}.$$

Thus we may choose $w \in U$ so that $\langle u, v, w\rangle \in F\{\langle P, g\rangle\}$. Since $c_1\{\langle P, g\rangle\} \cap d_{12} = \{\langle R, f_\epsilon\rangle\}$ and $d_{12} \cap c_1(c_0\{\langle P, g\rangle\} \cap d_{01}) = \{\langle R, f_\zeta\rangle\}$, it follows that $\{u, w\} \in T_\epsilon$ and $\{w, v\} \in T_\zeta$, as desired in (b).

(ii) $\Rightarrow$ (i). Assume (ii). For each $x \in {}^3U$ let

$$R_x = \{(i, j) \in {}^23 : x_i = x_j\},$$

and for any $(i, j) \in R_x'$ let $f_xij =$ the $\delta < \gamma$ such that $\{x_i, x_j\} \in T_\delta$. By (a), $\langle R_x, f_x\rangle \in \mathfrak{C}(3, 3, \gamma)$. Then for any $X \subseteq \mathfrak{C}(3, 3, \gamma)$ we set

$$FX = \{x \in {}^3U : \langle R_x, f_x\rangle \in X\}.$$

Clearly F is a Boolean homomorphism. To show that F is one-one, suppose that $0 \neq X \subseteq \mathfrak{C}(3, 3, \gamma)$; say $\langle S, g\rangle \in X$. If $S \neq$ identity on 3, obviously $x \in F\{\langle S, g\rangle\} \subseteq FX$ for some x. Suppose $S =$ identity on 3. Choose $\{u, v\} \in T_{g01}$. Then by (b) choose $w \neq u, v$ so that $\{u, w\} \in T_{g02}$ and $\{w, v\} \in T_{g12}$. Thus $\langle u, v, w\rangle \in F\{\langle S, g\rangle\} \subseteq FX$, as desired.

Clearly F preserves d_{ij} for any $i, j < 3$; also it is clear that $C_iFX \subseteq Fc_iX$ whenever $i < 3$ and $X \subseteq \mathfrak{C}(3, 3, \gamma)$. To show that

$Fc_iX \subseteq C_iFX$, suppose that $x \in Fc_iX$. Thus $\langle R_x, f_x\rangle \in c_iX$, so there is an $\langle S, g\rangle \in X$ such that $R_x \cap {}^3(\alpha \sim \{i\}) = S \cap {}^3(\alpha \sim \{i\})$ and $f_xkl = gkl$ if $k\,R'_x\,l$ and $k, l \neq i$. Let $3 = \{i, k, l\}$. If $S \neq$ identity on 3, it is easily seen that $x \in C_iFX$. Hence assume that $S =$ identity on 3. By (b), choose $w \in U \sim \{x_k, x_l\}$ so that $\{x_k, w\} \in T_{gik}$ and $\{w, x_l\} \in T_{gil}$. Then with

$$y = \{(k, x_k), (l, x_l), (i, w)\}$$

we have $\langle R_y, f_y\rangle = \langle S, g\rangle$ and hence $x \in C_iFX$.

This completes the proof.

THEOREM 4.6: *If $\alpha < \omega$, $\gamma \geqq \alpha - 1$, and $\alpha \geqq \beta \geqq 3$, then the following two conditions are equivalent:*

(i) *$\mathfrak{A}_{\alpha\beta\gamma}$ is representable;*

(ii) *there is a non-empty set U and a partition $\langle T_\delta : \delta < \gamma\rangle$ of S^2U such that*

(a) *no T_δ contains the complete graph on β vertices, i.e., if $V \subseteq U$* with $|V| = \beta$ then $S^2V \nsubseteq T_\delta$;

(b) *if $V \subseteq U$, $|V| < \alpha$, $f \in {}^V\gamma$, and there do not exist a $\delta < \gamma$ and a $W \subseteq V$* such that $|W| = \beta - 1$, $fw = \delta$ for all $w \in W$, and $S^2W \subseteq T_\delta$, *then there is a $u \in U \sim V$ such that $\{v, u\} \in fv$ for all $v \in V$.*

Proof: (i) $\Rightarrow$ (ii). By Corollary 4.4, let F be an isomorphism from $\mathfrak{A}_{\alpha\beta\gamma}$ onto a $\mathbf{Cs}^U$, where $U \neq 0$. Let R be the equivalence relation on α associated with the partition $\{\{0\}, \alpha \sim \{0\}\}$, and for each $\delta < \gamma$ let f_δ be the mapping of R' into γ such that $f_\delta 01 = \delta$. Then, as in the proof of 4.5,

(12) if $\quad \langle u, v, v, \cdots\rangle \in F\{\langle R, f_\delta\rangle\}$

then $\quad \langle v, u, u, \cdots\rangle \in F\{\langle R, f_\delta\rangle\}$.

Now for each $\delta < \gamma$ we let

$$T_\delta = \{\{u, v\} \in S^2U : \langle u, v, v, \cdots\rangle \in \{\langle R, f_\delta\rangle\}\}.$$

Clearly, then, $\langle T_\delta : \delta < \gamma\rangle$ is a partition of S^2U. To check (a), suppose $V \subseteq U$, $|V| = \beta$, and $S^2V \subseteq T_\delta$. Thus for any two distinct $u, v \in V$ we have $\langle u, v, v, \cdots \rangle \in F\{\langle R, f_\delta \rangle\}$. Let $w \in {}^\alpha U$ be such that for some $\epsilon \leqq \alpha$, $\epsilon \upharpoonright w$ is a one-one map onto V. For $\Gamma \subseteq \alpha$, say $\Gamma = \{\lambda_0, \cdots, \lambda_{k-1}\}$, let $c_{(\Gamma)} = c_{\lambda_0} \cdots c_{\lambda_{k-1}}$. Let

$$X = \bigcap_{0<i<\epsilon} c_{(\alpha \sim \{0,1\})}\{\langle R, f_\delta \rangle\}$$

$$\cap \bigcap_{0<i<j<\epsilon} c_0[d_{0i} \cap c_1(d_{1j} \cap c_{(\alpha \sim \{0,1\})}\{\langle R, f_\delta \rangle\}.$$

Then it is easily verified that $X = 0$, but on the other hand $w \in FX$. This contradiction shows that (a) holds.

Now to check (b), we assume its hypothesis. Let $x \in {}^\alpha U$ be such that for some $\epsilon < \alpha$, x maps ϵ one-one onto V. Let $x \in F\{\langle Q, g \rangle\}$. Now we set

$$S = [Q \cap {}^2(\alpha \sim \{\alpha - 1\})] \cup \{(\alpha - 1, \alpha - 1)\}.$$

Clearly S is an equivalence relation on α. For $k, l \in \alpha \sim \{\alpha - 1\}$ and $kQ'l$ we let $hkl = gkl$. For $i < \epsilon$ we let $h(\alpha - 1, i) = h(i, \alpha - 1) = fx_i$. Choose $j_1, \cdots, j_m \in (\alpha - 1) \sim \epsilon$ so that

$$j_s R' j_t \qquad \text{for} \qquad 1 \leqq s < t \leqq m;$$

if $k \in (\alpha - 1) \sim \epsilon$ then there is an s with $1 \leqq s \leqq m$ such that $k\, R\, j_s$.

Now let $h(j_1, \alpha - 1), h(j_2, \alpha - 1), \cdots, h(j_m, \alpha - 1)$ be distinct members of $\gamma \sim \{h(i, \alpha - 1):\ i < \epsilon\}$. For $k \in (\alpha - 1) \sim (\epsilon \cup \{j_s\colon 1 \leqq s \leqq m\})$ let s be such that $1 \leqq s \leqq m$ and $k\, R\, j_s$ (there is only one such s), and let $h(k, \alpha - 1) = h(j_s, \alpha - 1)$. Finally, if $l \in (\alpha - 1) \sim \epsilon$ we let $h(\alpha - 1, l) = h(l, \alpha - 1)$. This completes the definition of h; clearly h maps S' into γ. We claim that $\langle S, h \rangle \in \mathfrak{C}(\alpha, \beta, \gamma)$. Indeed, conditions (1)–(4) are clear from the definitions. Now, in order to check (5), suppose T is a selection set for S with $|T| = \beta$. If $\alpha - 1 \notin T$, the desired conclusion is obvious. Suppose $\alpha - 1 \in T$. If $k \in T$ for some k with $\epsilon \leqq k < \alpha - 1$, then again the desired conclusion is clear. Suppose there is no such k. Assume that $|\{hij\colon i, j \in T, i \neq j\}| = 1$; say $\{hij\colon i, j \in T, i \neq j\} = \{\delta\}$. Now from the fact that $x \in F\{\langle Q, g \rangle\}$ it easily follows that for distinct $i, j \in T \sim \{\alpha - 1\}$ we have $\{x_i, x_j\} \in T_{gij} = T_{hij} = T_\delta$. Furthermore, $fx_i = \delta$ for all $i \in T \sim \{\alpha - 1\}$. This contradicts the hypothesis of (b).

Hence our assumption that $|\{hij: i, j \in T, i \neq j\}| = 1$ is false. Thus $\langle S, h\rangle \in \mathcal{C}(\alpha, \beta, \gamma)$, as stated.

Now clearly $\langle Q, g\rangle \in c_{\alpha-1}\{\langle S, h\rangle\}$; since $x \in F\{\langle Q, g\rangle\}$, we may hence choose $u \in U$ such that $\langle x_0, \cdots, x_{\alpha-2}, u\rangle \in F\{\langle S, h\rangle\}$. Clearly u is as desired in the conclusion of (b).

(ii) $\Rightarrow$ (i). Assume (ii). For each $x \in {}^\alpha U$ let

$$R_x = \{(i, j) \in {}^2\alpha: x_i = x_j\},$$

and for any $(i, j) \in R'_x$ let $f_x ij =$ the $\delta < \gamma$ such that $\{x_i, x_j\} \in T_\delta$. By (a), $\langle R_x, f_x\rangle \in \mathcal{C}(\alpha, \beta, \gamma)$. Now for any $X \subseteq \mathcal{C}(\alpha, \beta, \gamma)$ we set

$$FX = \{x \in {}^\alpha U: \langle R_x, f_x\rangle \in X\}.$$

Clearly F is a Boolean homomorphism. To show that F is one-one, suppose that $0 \neq X \subseteq \mathcal{C}(\alpha, \beta, \gamma)$; say $\langle S, g\rangle \in X$. Choose $j_1, \cdots, j_m \in \alpha$ satisfying the following conditions:

(13) $\qquad j_s\, S'\, j_t \quad$ if $\quad 1 \leqq s < t \leqq m,$

(14) if $\; k \in \alpha, \;$ then $\; k\, S\, j_s \;$ for some s with $\; 1 \leqq s \leqq m.$

We now define $xj_1, \cdots, xj_m$ by recursion. Let xj_1 be any element of U. Now suppose that $1 \leqq s < m$ and $xj_1, \cdots, xj_s$ have been defined so that the following conditions hold:

(15) $\qquad xj_1, \cdots, xj_s \quad$ are all distinct;

(16) if $\quad 1 \leqq t < u \leqq s, \quad$ then $\quad \{xj_t, xj_u\} \in T_{gtu}.$

Let $V = \{xj_t: 1 \leqq t \leqq s\}$. Since $s < m \leqq \alpha$, $|V| < \alpha$. For each $xj_t \in V$ let $fxj_t = gj_t j_{s+1}$. Then the hypothesis of (b) holds. In fact, suppose there exist $\delta < \gamma$ and $W \subseteq V$ so that $|W| = \beta - 1$, $fw = \delta$ for all $w \in W$, and $S^2W \subseteq T_\delta$. Say $W = \{xj_t: t \in Z\}$ where $Z \subseteq \{t: 1 \leqq t \leqq s\}$. Thus by (16), $gj_t j_u = \delta$ for distinct $t, u \in Z$, and $gj_t j_{s+1} = \delta$ for any $t \in Z$. This contradicts condition (5) for $\langle S, g\rangle$. Thus, indeed, the hypothesis of (b) holds. By the conclusion of (b), let $xj_{s+1} \in U \sim V$ be such that $\{xj_t, xj_{s+1}\} \in fxj_t$ for all t with $1 \leqq t \leqq s$. Thus (15) and (16) hold for $s + 1$. This completes the construction of $xj_1, \cdots, xj_m$. If $k \in \alpha \sim \{j_1, \cdots, j_m\}$, let s be such that $1 \leqq s \leqq m$ and $k\, S\, j_s$ (by (14), where s is unique by (13)), and set $xk = xj_s$. Thus $x \in {}^\alpha U$. Clearly $x \in F\{\langle S, g\rangle\} \subseteq FX$, as desired. Thus F is one-one.

Clearly F preserves d_{ij} for any $i, j < \alpha$; also it is clear that $C_iFX \subseteq Fc_iX$ whenever $i < \alpha$ and $X \subseteq \mathfrak{C}(\alpha, \beta, \gamma)$. To show that $Fc_iX \subseteq C_iFX$, suppose that $x \in Fc_iX$. Thus $\langle R_x, f_x\rangle \in c_iX$, so there is an $\langle S, g\rangle \in X$ such that $R_x \cap {}^3(\alpha \sim \{i\}) = S \cap {}^3(\alpha \sim \{i\})$ and $f_xkl = gkl$ if $k\, R_x'\, l$ and $k, l \neq i$. If $i\, S\, j$ for some $j \neq i$, it is clear that $x \in C_iFX$. Hence assume that $i\, S'\, j$ for all $j \neq i$. Choose $j_1, \cdots, j_m \in \alpha \sim \{i\}$ such that $j_s\, S'\, j_t$ if $1 \leqq s < t \leqq m$;

$$\text{if} \quad k \in \alpha \sim \{i\}, \quad \text{then} \quad k\, S\, j_s \quad \text{for some } s \text{ with } 1 \leqq s \leqq m.$$

Let $V = \{xj_s\colon 1 \leqq s \leqq m\}$. For each s with $1 \leqq s \leqq m$ let $hxj_s = gij_s$. Then the hypothesis of (b) holds, as is easily seen. Applying (b), we get a $u \in U \sim V$ such that $\{xj_s, u\} \in hxj_s$ for each s with $1 \leqq s \leqq m$. Let y be like x except that $y_i = u$. Clearly $R_y = S$ and $f_y = g$. It follows that $x \in C_iFX$, as desired.

The combinatorial condition expressed in 4.6(ii) may be loosely termed a *free decomposition of the complete graph on U into subgraphs each excluding the complete graph on β vertices.* This condition has not been investigated in the literature, as far as this author knows. However, condition 4.6(ii) (a) by itself has been extensively investigated. The basic result here is the following theorem of Ramsey (see Ramsey [15]):

THEOREM 4.7: *Suppose $3 \leqq \beta < \omega$ and $2 \leqq \gamma < \omega$. Then there is an integer $n(\beta, \gamma) \in \omega \sim 1$ with the following property. If U is a set with at least $n(\beta, \gamma)$ elements and if $\langle T_\delta\colon \delta < \gamma\rangle$ is a partition of S^2U, then there exist a $\delta < \gamma$ and a subset V of U with $|V| = \beta$ such that $S^2V \subseteq T_\delta$.*

From this theorem and the proof of 4.6 it follows that if $\mathfrak{A}_{\alpha\beta\gamma}$ is representable, say isomorphic to a $\boldsymbol{Cs}^U$, then $|U| < n(\beta, \gamma)$. Thus a good knowledge of representation of the algebras $\mathfrak{A}_{\alpha\beta\gamma}$ would yield lower bounds for the Ramsey numbers $n(\beta, \gamma)$. We do not have such knowledge yet. However, in [3] it is shown directly that $\mathfrak{A}(3, 3, 2)$ and $\mathfrak{A}(3, 3, 3)$ are representable; but since $n(3, 2) = 6$ and $n(3, 3) = 17$ by [4], no new information on the Ramsey numbers is obtained.

On the other hand, $|A_{\alpha\beta\gamma}|$ forces a lower bound on $|U|$. We shall give such a lower bound for the case $\beta = 3$; for the proof see the proof of Theorem 1.8 of [13].

THEOREM 4.8: *If* $\mathfrak{A}_{\alpha 3\gamma}$ *is isomorphic to a* $\mathbf{Cs}^U$ *and* $\alpha < \omega$, $2 \leqq \gamma < \omega$, *then*

$$|S^2U| \geqq (\gamma - 2)^{2\alpha-6}.$$

Now upper bounds for the Ramsey numbers $n(\beta, \gamma)$ are not known in general, but for $\beta = 3$ we have the following result of Greenwood–Gleason [4]:

THEOREM 4.9: $n(3, \gamma) \leqq [\gamma!\, e] + 1$.

COROLLARY 4.10: *If* $\mathfrak{A}(\alpha, 3, \gamma)$ *is representable, then*

$$(\gamma - 2)^{2\alpha-6} \leqq ([\gamma!\, e] + 1)^2 \leqq 9 \cdot \gamma!^2.$$

Corollary 4.10 implies that certain of the algebras $\mathfrak{A}(\alpha, 3, \gamma)$ are nonrepresentable. For example, a simple computation shows that $\mathfrak{A}(13, 3, 13)$ is non-representable. In [13] these methods are refined so as to yield a far-reaching generalization of the result mentioned at the end of section 3. Namely, with the essential use of Theorem 1.3, it is shown there that $\boldsymbol{R}_\alpha$ is not finitely axiomatizable for any $\alpha \geqq 3$.

REFERENCES

1. Bruck, R. H., *A Survey of Binary Systems.* Springer-Verlag, 1966.
2. Bruck, R. H. and H. J. Ryser, "The nonexistence of certain finite projective planes," *Canad. J. Math.*, **1** (1949), 88–93.
3. Demaree, D. B., "Studies in algebraic logic," Doctoral dissertation, University of California, Berkeley 1970.
4. Greenwood, R. E. and A. M. Gleason, "Combinatorial relations and chromatic graphs," *Canad. J. Math.*, **7** (1955), 1–7.
5. Halmos, P. R., *Algebraic Logic.* Chelsea, 1962.
6. Henkin, L., J. D. Monk, and A. Tarski, *Cylindric Algebras,* Part I. North-Holland, 1971.

7. Johnson, J. S., "Nonfinitizability of classes of representable polyadic algebras," *J. Symbolic Logic*, **34** (1969), 344–352.

8. Jónsson, D., "Representation of modular lattices and of relation algebras," *Trans. Amer. Math. Soc.*, **92** (1959), 449-464.

9. Lyndon, R. C., "Relation algebras and projective geometries," *Michigan Math. J.*, **8** (1961), 21–28.

10. McKenzie, R., "Representations of integral relation algebras," *Michigan Math. J.*, **17** (1970), 279–287.

11. Monk, J. D., "On representable relation algebras," *Michigan Math. J.*, **11** (1964), 207–210.

12. ———, "Model-theoretic methods and results in the theory of cylindric algebras," *The Theory of Models*. North-Holland, 1965, 238–250.

13. ———, "Nonfinitizability of classes of representable cylindric algebras," *J. Symbolic Logic*, **34** (1969), 331–343.

14. ———, "On an algebra of sets of finite sequences," *J. Symbolic Logic*, **35** (1970), 19–28.

15. Ramsey, F. P., "On a problem of formal logic," *Proc. London. Math. Soc.*, **30** (1930), 264–286.

16. Seidenberg, A., *Lectures in Projective Geometry*. Van Nostrand, 1962.

POST ALGEBRAS AS A SEMANTIC FOUNDATION OF *m*-VALUED LOGICS

Helena Rasiowa

INTRODUCTION

Remarkable advances in universal algebra have had a strong effect on the various branches of mathematics. In mathematical logic, too, algebraic methods have found many applications.

The study of the relationships between logic and algebra was initiated by Boole[1], who gave birth to modern logic. Boole's idea was to look for analogies between the laws of logic and those of algebra. It was later continued in the works of Jevons, Peirce and Schröder.

The works of Cantor (published mainly from 1871 to 1883), which developed set theory into a separate branch of mathe-

[1] Cf. Boole, G., "The mathematical analysis of logic . . .", 1847 (repr. Oxford and New York, 1948), "An investigation of the laws of thought . . .", 1854 (repr. New York, 1951).

matics, and those of Frege (1884, 1891, 1903), which were intended to lay logical foundations of arithmetic, accounted for a reorientation of logical researches. The turn of the 19th century saw the foundations of logic and mathematics shaken by the discovery of certain semantic and set-theoretical antinomies. A broad discussion of the methods which would make it possible to avoid antinomies in logic and mathematics gave rise to three different approaches: Russell's logicism, Hilbert's formalism and Brouwer's intuitionism[2]. The principal idea of intuitionism was to link the concept of existence with the possibility of constructing an object which is claimed to exist. This idea resulted in the rejection by Brouwer of those laws of classical logic which may lead to ineffective proofs of existence. This yielded a new, intuitionistic logic, which is a part of classical logic. Intuitionistic logic later came to be formalized by Heyting [12]. Brouwer was not the only mathematician to be critical of classical logic. Criticism resulted in the emergence of other systems of logic, such as the modal logics of Lewis [17], many-valued logics (Łukasiewicz [20], [21], and Post [32]), positive logic (Hilbert and Bernays [14]), minimal logic (Johansson [16]), constructive logic with strong negation (Nelson [31] and Markov [25]), etc.

The birth of non-classical logics anew stimulated investigations into relationships between these logics and the corresponding classes of abstract algebras. The years 1920–40 saw, in particular in the Polish school of logic, researches on non-classical propositional calculi conducted by what is termed the logical matrix method. Since logical matrices are certain abstract algebras, this led to the use of the algebraic method in logic. A description of the general algebraic method for the propositional calculi, illustrated by numerous examples of non-classical logics, is to be found in *Algebraic Approach to Non-classical Logics* by the present author, *Studies in Logic and the Foundations of Mathematics*, Amsterdam: North-Holland, Warszawa: PWN, 1974.

---

[2] Cf. Heyting, A., [13].

1. ALGEBRAIZATION OF VARIOUS PROPOSITIONAL CALCULI

It is common knowledge that classical logic has an algebraic interpretation in Boolean algebras. The class of all Boolean algebras is equationally definable by means of a system of axioms[3], which will be omitted here. Roughly speaking, every Boolean algebra is a non-empty set $\boldsymbol{A}$ with the operations of join ($\cup$), meet ($\cap$), and complementation ($-$)[4], which follow laws analogous to those governing the set-theoretical operations of union, intersection, and complementation, respectively. It is convenient, with a view to applications in logic, to introduce one more operation, termed codifference, denoted by $\Rightarrow$, and defined thus: $a \Rightarrow b = -a \cup b$. A partial order relation $\leq$ (i.e., reflexive, transitive and antisymmetric) can be defined thus:

$$(1) \qquad a \leq b \quad \text{iff} \quad a \cap b = a,$$

in each Boolean algebra. Its analogue in the algebra of sets is the inclusion relation. Each Boolean algebra has two designated elements, namely the unit element $\vee$ such that, for every element x in that algebra, $x \leq \vee$, and the zero element $\wedge$ such that, for every element x in that algebra, $\wedge \leq x$. The relation $\leq$ can also be defined by the equivalence: $a \leq b$ iff $a \Rightarrow b = \vee$. Examples of Boolean algebras can be seen in fields of sets, i.e., non-empty families $\boldsymbol{B}(X)$ of subsets of any space X, such that the union and the intersection of any two sets in $\boldsymbol{B}(X)$ is also in $\boldsymbol{B}(X)$, and the complement of any set in $\boldsymbol{B}(X)$ is also in $\boldsymbol{B}(X)$. By the Stone representation theorem [50] each Boolean algebra is isomorphic with a field of sets. The simplest example of a Boolean algebra is provided by the two-element Boolean algebra $\boldsymbol{B}_0$, whose set of elements consists of $\wedge$ and $\vee$ and which is isomorphic with the field of subsets of an arbitrary space X, consisting of the empty set O and the whole space X.

A mapping h of the set of elements of a Boolean algebra $\boldsymbol{A}$ onto

---

[3] A system of axioms is to be found, e.g., in [41].

[4] For convenience we shall not strictly distinguish between an abstract algebra and the set of its elements.

the set of elements of a Boolean algebra $\boldsymbol{B}$ is called an epimorphism of $\boldsymbol{A}$ onto $\boldsymbol{B}$ if $h(a \cup b) = h(a) \cup h(b)$, $h(a \cap b) = h(a) \cap h(b)$, and $h(-a) = -h(a)$. The inverse image of the unit element by an epimorphism h, i.e., $h^{-1}(\vee)$, is called the kernel of h. The kernels of epimorphisms of Boolean algebras are filters, i.e., non-empty subsets ∇ of the set of all elements of a given Boolean algebra, such that $a \cap b \in \nabla$ iff $a \in \nabla$ and $b \in \nabla$. This property is equivalent to the following one: if $a, b \in \nabla$, then $a \cap b \in \nabla$, and if $a \in \nabla$, then for each b such that $a \leq b$, $b \in \nabla$. Another equivalent definition of a filter is as follows: ∇ is a filter if $\vee \in \nabla$ and if the condition $a, a \Rightarrow b \in \nabla$ implies $b \in \nabla$. Every filter ∇ in a Boolean algebra $\boldsymbol{A}$ determines a congruence relation $\approx_\nabla$, which is defined thus: $a \approx_\nabla b$ iff $a \Rightarrow b \in \nabla$ and $b \Rightarrow a \in \nabla$. The quotient algebra $\boldsymbol{A}/\approx_\nabla$ (i.e., formed of all equivalence classes under this relation), also denoted by $\boldsymbol{A}/\nabla$, is a Boolean algebra and the mapping $h(a) = [a]$, where $[a]$ denotes the equivalence class under $\approx_\nabla$, determined by $a \in \boldsymbol{A}$, is an epimorphism of $\boldsymbol{A}$ onto $\boldsymbol{A}/\nabla$. Moreover, ∇ is the kernel of h. Thus the kernels of the epimorphisms of a Boolean algebra coincide with the filters of this algebra. It can easily be deduced that, for each element a of a Boolean algebra $\boldsymbol{A}$ and a filter ∇,

$$(2) \qquad a \in \nabla \quad \text{iff} \quad [a] = \vee \quad \text{in} \quad \boldsymbol{A}/\nabla.$$

What is called a prime filter is the kernel of an epimorphism onto the two-element Boolean algebra $\boldsymbol{B}_0$. Thus, for each prime filter ∇ in a Boolean algebra $\boldsymbol{A}$, the quotient algebra $\boldsymbol{A}/\nabla$ is the two-element Boolean algebra. Prime filters are proper filters, i.e., different from the whole algebra in question, such that

$$(3) \qquad a \cup b \in \nabla \quad \text{iff either} \quad a \in \nabla \quad \text{or} \quad b \in \nabla.$$

Another characteristic property of prime filters in Boolean algebras is this: for each element a of a given algebra, exactly one of the elements a and $-a$ is in ∇. It can easily be proved, by reference to the Kuratowski–Zorn lemma, that

(4) for each element $a \neq \wedge$ of any Boolean algebra $\boldsymbol{A}$ there is a prime filter ∇ such that $a \in \nabla$,

and also that

(5) each proper filter in a Boolean algebra is contained in a prime filter.

Consider now a formalized language $\mathfrak{L}_\kappa$ of the classical (two-valued) propositional calculus. The formulas (meaningful expressions) of that language are constructed of propositional variables $p_1, p_2, \cdots$ by means of the logical operations of disjunction ($\cup$), conjunction ($\cap$), implication ($\Rightarrow$), negation ($-$)[5], and brackets. For instance, $(-(p_1 \cap p_2) \Rightarrow (-p_1 \cup -p_2))$ is an example of a formula. Each formula α of $\mathfrak{L}_\kappa$ can be interpreted in any Boolean algebra $\boldsymbol{A}$ as an algebraic function $\alpha_{\boldsymbol{A}}$ of that algebra if the propositional variables in α are treated as variables ranging over $\boldsymbol{A}$, and the logical operations $\cup$, $\cap$, $\Rightarrow$, $-$, as the corresponding algebraic operations in $\boldsymbol{A}$. In particular, if the formulas of $\mathfrak{L}_\kappa$ are interpreted in the two-element Boolean algebra $\boldsymbol{B}_0$ and if the symbol $\vee$ is treated as the symbol of any true proposition, and the symbol $\wedge$ is treated as the symbol of any false proposition, then the condition: $\alpha_{\boldsymbol{B}_0} = \vee$ identically, states that α is a true proposition regardless of whether those propositions of which α is constructed be true or false. If this condition is satisfied, then it is said that α is a law (tautology) of the classical propositional calculus. The opinion that any proposition can be only true or false is one of the foundations of classical logic.

Every function v which assigns elements of a Boolean algebra $\boldsymbol{A}$ to the propositional variables which occur in $\mathfrak{L}_\kappa$ is called a valuation of $\mathfrak{L}_\kappa$ in $\boldsymbol{A}$. For every formula α, $\alpha_{\boldsymbol{A}}$ may, of course, be treated as a mapping $\alpha_{\boldsymbol{A}}\colon V \rightarrow \boldsymbol{B}$, where V is the set of the propositional variables that occur in $\mathfrak{L}_\kappa$. Given any set $\mathfrak{A}$ of formulas of $\mathfrak{L}_\kappa$ we say that a formula α is a consequence of $\mathfrak{A}$, if, for every valuation $v\colon V \rightarrow \boldsymbol{B}_0$ of $\mathfrak{L}_\kappa$ in the two-element Boolean algebra $\boldsymbol{B}_0$ such that $\beta_{\boldsymbol{B}_0}(v) = \vee$ for every $\beta \in \mathfrak{A}$, we have $\alpha_{\boldsymbol{B}_0}(v) = \vee$. The intuitive

---

[5] We use the same symbols $\cup$, $\cap$, $\Rightarrow$, $-$, to stand for set operations, algebraic operations and the corresponding propositional operations in order to emphasize the close reciprocal relationship between them. Other similar situations will be dealt with in the same manner.

meaning of the statement that α is a consequence of $\mathfrak{A}$ is clear: if all formulas β in $\mathfrak{A}$, for fixed logical values of all propositional variables, stand for true propositions, then α also stands for a true proposition. Partial k-argument functions r ($k = 1, 2, \cdots$) from the set of all formulas into itself such that, if $r(\alpha_1, \cdots, \alpha_k) = \beta$ then β is a consequence of the set $\{\alpha_1, \cdots, \alpha_k\}$ of formulas, are called rules of inference.

Axiomatization of the classical propositional calculus consists in selecting, from among the laws of that calculus, a recursive set[6] $\mathfrak{A}_l$ of formulas (a set of logical axioms) and a finite set of rules of inference so that the following condition is satisfied: every formula α which is a law of the classical propositional calculus can be derived from logical axioms by the application of rules of inference (in a finite number of steps). This is written "$\vdash\alpha$" and reads "α is a theorem".

The following set of formulas is an example of a set $\mathfrak{A}_l$ of logical axioms of the classical propositional calculus:

$$
\begin{array}{ll}
(A_1) & (\alpha \Rightarrow (\beta \Rightarrow \alpha)),\\
(A_2) & ((\alpha \Rightarrow (\beta \Rightarrow \gamma)) \Rightarrow ((\alpha \Rightarrow \beta) \Rightarrow (\alpha \Rightarrow \gamma))),\\
(A_3) & (\alpha \Rightarrow (\alpha \cup \beta)),\\
(A_4) & (\beta \Rightarrow (\alpha \cup \beta)),\\
(A_5) & ((\alpha \Rightarrow \gamma) \Rightarrow ((\beta \Rightarrow \gamma) \Rightarrow ((\alpha \cup \beta) \Rightarrow \gamma))),\\
(A_6) & ((\alpha \cap \beta) \Rightarrow \alpha),\\
(A_7) & ((\alpha \cap \beta) \Rightarrow \beta),\\
(A_8) & ((\gamma \Rightarrow \alpha) \Rightarrow ((\gamma \Rightarrow \beta) \Rightarrow (\gamma \Rightarrow (\alpha \cap \beta)))),\\
(A_9) & ((\alpha \Rightarrow -\beta) \Rightarrow (\beta \Rightarrow -\alpha)),\\
(A_{10}) & (-(\alpha \Rightarrow \alpha) \Rightarrow \beta),\\
(A_{11}) & (\alpha \cup -\alpha),
\end{array}
$$

where α, β, γ are any formulas. In this case one rule of inference, called modus ponens, is adopted. This rule assigns the formula β to each pair of formulas of the form: α and $(\alpha \Rightarrow \beta)$.

Let S_κ be the formalized system of the classical propositional

---

[6] Roughly speaking this means that there is an effective method which enables us to decide in a finite number of steps whether a formula α belongs to $\mathfrak{A}$ or not.

calculus based on the above set $\mathfrak{A}_l$ of logical axioms and the rule of modus ponens. The following well-known theorem holds:

THEOREM 1: *For every formula α in the classical propositional calculus $\mathcal{S}_\kappa$ the following conditions are equivalent:*

(i) $\vdash \alpha$,
(ii) $\alpha_{\boldsymbol{B}} = \vee$ *identically in every Boolean algebra* $\boldsymbol{B}$,
(iii) $\alpha_{\boldsymbol{B}(X)} = X$ *identically in every field of sets* $\boldsymbol{B}(X)$,
(iv) $\alpha_{\boldsymbol{B}_0} = \vee$ *identically in the two-element Boolean algebra* $\boldsymbol{B}_0$.

The equivalence of (i) and (iv) is the completeness theorem for $\mathcal{S}_\kappa$. Proofs of the implications (i) $\rightarrow$ (ii), (ii) $\rightarrow$ (iii), (iii) $\rightarrow$ (iv) do not present any difficulties. The proof of (iv) $\rightarrow$ (i) is based on the theorem (4) on the existence of prime filters in Boolean algebras. The idea of the proof refers to the Lindenbaum–Tarski method of constructing, for formal systems, corresponding abstract algebras consisting of formulas of those systems. The method is a point of departure for numerous applications of algebra in logic. Let us consider the set of all formulas of $\mathcal{S}_\kappa$ with logical operations as an abstract algebra similar to Boolean algebras. The relation $\approx$, which holds between any two formulas α, β iff $\vdash(\alpha \Rightarrow \beta)$ and $\vdash(\beta \Rightarrow \alpha)$, is a congruence in that algebra. The quotient algebra, denoted by $\boldsymbol{A}(\mathcal{S}_\kappa)$, is a Boolean algebra called the algebra of $\mathcal{S}_\kappa$. Let $\| \alpha \|$ stand for that element of this algebra which is determined by α. It is proved that

$$\| \alpha \| = \vee \quad \text{iff} \quad \vdash \alpha. \tag{6}$$

Hence if $\vdash \alpha$ does not hold, then $\| \alpha \| \neq \vee$ and $-\| \alpha \| \neq \wedge$. By (4), there is a prime filter ∇ such that $-\| \alpha \| \in \nabla$. The quotient algebra $\boldsymbol{A}(\mathcal{S}_\kappa)/\nabla$ is the two-element Boolean algebra and by (2), its element $-[\| \alpha \|] = [-\| \alpha \|] = \vee$. Hence $[\| \alpha \|] = \wedge$. For the valuation v_0 in $\boldsymbol{A}(\mathcal{S}_\kappa)/\nabla$ which to each variable p_i, $i = 1, 2, \cdots$, assigns the element $[\| p_i \|]$ we have, for each formula β, $\beta_{\boldsymbol{A}(\mathcal{S}_\kappa)/\nabla}(v_0) = [\| \beta \|]$. In particular, $\alpha_{\boldsymbol{A}(\mathcal{S}_\kappa)/\nabla}(v_0) = [\| \alpha \|] = \wedge$. This shows that (iv) implies (i).

It is worth mentioning that the two-element Boolean algebra $\boldsymbol{B}_0 = (\{\wedge, \vee\}, \cup, \cap, \Rightarrow, -)$ is functionally complete, which means that for every $n = 0, 1, 2, \cdots$, every function $f: \{\wedge, \vee\}^n \to \{\wedge, \vee\}$ can be defined by the operations $\cup, \cap, \Rightarrow, -$.

A system $\mathcal{S}_\chi$ of the intuitionistic propositional calculus can be obtained from $\mathcal{S}_\kappa$ by a weakening of the system of logical axioms, namely by dropping (A_{11}), and by replacing the negation sign— by $\urcorner$[7]. The formulas of the intuitionistic propositional calculus can also be interpreted algebraically in abstract algebras of the type $\boldsymbol{A} = (A, \vee, \cup, \cap, \Rightarrow, \urcorner)$ with a designated element $\vee$; the problem then is whether there is a class $\boldsymbol{K}_\chi$ of similar algebras which have the following properties:

(i) α is a theorem of $\mathcal{S}_\chi$ iff $\alpha_{\boldsymbol{A}} = \vee$ identically in each $\boldsymbol{A} \in \boldsymbol{K}_\chi$,
(ii) *modus ponens* is preserved in algebras of $\boldsymbol{K}_\chi$, which means that

$$\text{if} \quad \alpha_{\boldsymbol{A}}(v) = \vee \quad \text{and} \quad (\alpha \Rightarrow \beta)_{\boldsymbol{A}}(v) = \vee,$$
$$\text{then} \quad \beta_{\boldsymbol{A}}(v) = \vee \quad \text{in each} \quad \boldsymbol{A} \in \boldsymbol{K}_\chi,$$

and whether there is in that class a finite algebra $\boldsymbol{A}_0$ such that, for every formula α, α is a theorem of $\mathcal{S}_\chi$ iff $\alpha_{\boldsymbol{A}_0} = \vee$ identically.

Research conducted by Tarski [52], McKinsey and Tarski [27], [28], and also by Stone [51] has shown the class of pseudo-Boolean algebras to have these properties. The class of those algebras is definable equationally by a finite system of axioms[8], which will not be given here. Examples of such algebras can be obtained if we consider any topological space X (in the sense of Kuratowski's axiom system) with an interior operation $\boldsymbol{I}$ and the family $\boldsymbol{G}(X)$ of all open subsets of that space. The algebras $(\boldsymbol{G}(X), X, \cup, \cap, \Rightarrow, \urcorner)$ with the set-theoretical operations $\cup, \cap$ and the operations $\Rightarrow, \urcorner$ defined by the equations $Y \Rightarrow Z = \boldsymbol{I}((X - Y) \cup Z)$, $\urcorner Y = \boldsymbol{I}(X - Y)$, and their subalgebras are pseudo-Boolean algebras called pseudo-fields of sets. By Stone's representation theorem [51] these cases cover all pseudo-Boolean algebras up to isomorphism. This yields the following theorem:

---

[7] In intuitionistic logic usually the symbol $\urcorner$ is adopted as the intuitionistic negation sign and this is the only reason of the replacement of $-$ by $\urcorner$.

[8] A system of axioms is to be found, e.g., in [41].

THEOREM 2: *The following conditions are equivalent, for every formula α, in the system $\mathcal{S}_\chi$ of the intuitionistic propositional calculus:*

(i) $\vdash \alpha$ (*α is a theorem in $\mathcal{S}_\chi$*),
(ii) $\alpha_{\boldsymbol{A}} = \vee$ *identically in every pseudo-Boolean algebra* $\boldsymbol{A}$,
(iii) $\alpha_{\boldsymbol{G}(X)} = X$ *identically in every pseudo-field* $\boldsymbol{G}(X)$ *of sets.*

K. Gödel has proved that the answer to the second question is in the negative. There is no finite algebra that would be characteristic of $\mathcal{S}_\chi$, contrary to the classical propositional calculus, of which the two-element Boolean algebra is characteristic. On the other hand McKinsey and Tarski [28] proved that each of the conditions (i)–(iii) in Theorem 2 is equivalent to each of the following two conditions:

(iv) $\alpha_{\boldsymbol{G}(X)} = X$ identically in the pseudo-field $\boldsymbol{G}(X)$ of all open subsets of a fixed normal topological space X which is dense in itself, such as, for instance, a Euclidean space;
(v) $\alpha_{\boldsymbol{A}} = \vee$ identically in every finite pseudo-Boolean algebra $\boldsymbol{A}$ whose number of elements is not greater than 2^{2^r}, where r stands for the number of all formulas occurring in α (sub-formulas of α).

It can easily be seen, from Theorem 2, that the formula $((p_1 \Rightarrow p_2) \cup (p_2 \Rightarrow p_1))$ is not a theorem in $\mathcal{S}_\chi$. By adding to the system of the logical axioms of $\mathcal{S}_\chi$ a new axiom scheme (A'_{11}) $((\alpha \Rightarrow \beta) \cup (\beta \Rightarrow \alpha))$ for all formulas α, β, we obtain a system $\mathcal{S}_{\chi l}$ of the linear intuitionistic logic (Dummet [6]), which is richer than $\mathcal{S}_\chi$ but poorer than $\mathcal{S}_\chi$. The class $\boldsymbol{K}_{\chi l}$ of algebras which characterizes $\mathcal{S}_{\chi l}$ is the class of all those pseudo-Boolean algebras in which the equation $(a \Rightarrow b) \cup (b \Rightarrow a) = \vee$ holds identically. Such algebras will be called linear pseudo-Boolean algebras. They include algebras of the type $(\boldsymbol{L}, \vee, \cup, \cap, \Rightarrow, \neg)$, where $\boldsymbol{L}$ is a linearly ordered set (a chain) with a greatest element $\vee$ and a least element $\wedge$. In this case $a \cup b = \max(a, b)$, $a \cap b = \min(a, b)$, $a \Rightarrow b = \vee$ if $a \leq b$ and $a \Rightarrow b = b$ for $b \leq a$, $b \neq a$, $\neg a = a \Rightarrow \wedge$ (i.e., $\neg a = \vee$ if $a = \wedge$ and $\neg a = \wedge$ if $a \neq \wedge$). Dummet [6] proved the following theorem:

THEOREM 3: *For every formula* α *in* $S_{\chi l}$ *the following conditions are equivalent:*

(i) $\vdash\alpha$ (α *is a theorem of* $S_{\chi l}$),

(ii) $\alpha_A = \vee$ *identically in every linear pseudo-Boolean algebra* $\boldsymbol{A}$,

(iii) $\alpha_L = \vee$ *identically in every chain* $\boldsymbol{L}$ *which is a pseudo-Boolean algebra,*

(iv) $\alpha_L = \vee$ *identically in some infinite chain* $\boldsymbol{L}$ *which is a pseudo-Boolean algebra,*

(v) $\alpha_L = \vee$ *identically in every finite chain* $\boldsymbol{L}$ *which is a pseudo-Boolean algebra,*

(vi) $\alpha_L = \vee$ *identically in every chain* $\boldsymbol{L}$ *with at most* $n + 2$ *elements, which is a pseudo-Boolean algebra, where* n *is the number of variables in* α.

Research on the algebraic interpretation of the propositional calculi in other known non-classical logics have resulted in singling out various interesting classes of algebras, some of which only will be mentioned here.

A system S_μ of the minimal propositional calculus can be obtained from S_χ by dropping the axiom scheme (A_{10}) from the system of logical axioms of S_χ. This propositional calculus is algebraically characterized by the class of μ-algebras (Rasiowa and Sikorski [40]). Examples of such algebras are provided by algebras $(\boldsymbol{G}(X), X, \cup, \cap, \Rightarrow, \neg)$ of all open subsets of any topological space X, with the set-theoretical operations $\cup$, $\cap$, the operation $\Rightarrow$ defined as in pseudo-fields of sets, and the operation $\neg$ defined by the equation $\neg Y = Y \Rightarrow \neg X$, where $\neg X$ is any fixed set in $\boldsymbol{G}(X)$. These algebras and their subalgebras cover all μ-algebras up to isomorphism (Rasiowa and Sikorski [40]).

A system S_π of the positive propositional calculus can be obtained from S_μ by omitting the axiom scheme (A_9) from the system of logical axioms of S_μ and by dropping the negation sign $\neg$ from the formalized language of S_μ. Propositional calculus S_π has its algebraic interpretation in the class of relatively pseudo-complemented lattices (Rasiowa and Sikorski [40]). All examples of these algebras are provided up to isomorphism by algebras

$(\boldsymbol{G}(X), X, \cup, \cap, \Rightarrow)$ of all open and all simultaneously open and dense subsets of any topological space X, where the operations $\cup, \cap, \Rightarrow$, are defined as in the pseudo-fields of sets, and by their subalgebras (Rasiowa and Sikorski [40]).

Those modal propositional calculi which correspond to the Lewis system S4 (Lewis and Langford [17]) have their algebraic interpretation in topological Boolean algebras. Examples of such algebras are provided by the algebras of all subsets of any topological space X with the set-theoretical operations $\cup, \cap, -$, the operation $\Rightarrow$ defined by the equation $Y \Rightarrow Z = (X - Y) \cup Z$, and the interior and closure operations that correspond, respectively, to the logical operations of necessity and possibility that occur in that logic. These algebras and their subalgebras cover all topological Boolean algebras up to isomorphism (McKinsey and Tarski [26]). Theorems analogous to Theorem 2 are also obtained here (McKinsey and Tarski [28]) and in the cases mentioned above.

The propositional calculi of the constructive logic with strong negation have an algebraic interpretation in the class of N-lattices (Rasiowa [36], Białynicki-Birula and Rasiowa [1]). The propositional calculi of Łukasiewicz's m-valued logics have an interpretation in abstract algebras called Łukasiewicz's algebras (cf. Moisil [29]). The structure of N-lattices and Łukasiewicz's algebras will not be discussed here.

2. POST ALGEBRAS AND m-VALUED PROPOSITIONAL CALCULI

Adoption of the principle that there are propositions which are neither true nor false, i.e., the principle that logical values other than those of truth and falsehood may exist, is the foundation of Łukasiewicz's many-valued logics (Łukasiewicz [20], [21], see also Łukasiewicz and Tarski [22]). Independently of Łukasiewicz, Post [32] has also constructed his m-valued logics. Post's and Łukasiewicz's many-valued propositional calculi were first constructed, as was also the case of the two-valued propositional calculi, by the truth-table matrix method, and not as formalized axiomatic systems. This method will be described below.

Let $e_0 \leq e_1 \leq \cdots \leq e_{m-1}$ be a chain, and let $\boldsymbol{P}_m = \{e_0, \cdots, e_{m-1}\}$. $\boldsymbol{P}_m$ will be treated as the set of all admissible logical values, where $e_0 = \wedge$ is the logical value of falsehood, and $e_{m-1} = \vee$, that of truth. We introduce algebraic operations $\cup, \cap, \Rightarrow, \neg$ in $\boldsymbol{P}_m$ which are defined as follows:

$$e_i \cup e_j = e_{\max(i,j)}; \qquad e_i \cap e_j = e_{\min(i,j)};$$

$$e_i \Rightarrow e_j = \begin{cases} e_{m-1} & \text{if } i \leqq j \\ e_j & \text{if } i > j \end{cases}; \qquad \neg e_i = e_i \Rightarrow e_0 = \begin{cases} e_{m-1} & \text{if } i = 0 \\ e_0 & \text{if } i \neq 0. \end{cases}$$

Observe that the algebra $(\boldsymbol{P}_m, \vee, \cup, \cap, \Rightarrow, \neg)$ is then a linear pseudo-Boolean algebra of m elements which is a chain (cf. Sec. 1, p. 100). For $m = 2$ we get the two-element Boolean algebra. For $m > 2$, contrary to the two-element Boolean algebra, this algebra is not functionally complete. To make it such we add new operations of one argument each, namely $D_1, \cdots, D_{m-1}$. These are defined thus:

$$D_i(e_j) = \begin{cases} e_{m-1} & \text{if} \quad i \leqq j \\ e_0 & \text{if} \quad i > j \end{cases}.$$

Note that for $m = 2$, the only new operation D_1 is the identity operation and therefore can be omitted.

The algebra $\boldsymbol{P}_m = (\boldsymbol{P}_m, \vee, \cup, \cap, \Rightarrow, \neg, D_1, \cdots, D_{m-1}, e_0, \cdots, e_{m-1})$, $m \geqq 2$, is functionally complete, i.e., every mapping $f\colon \boldsymbol{P}_m^k \to \boldsymbol{P}_m$, for $k = 0, 1, 2, \cdots$, can be presented as a composition of operations $\cup, \cap, \Rightarrow, \neg, D_1, \cdots, D_{m-1}, e_0, \cdots, e_{m-1}$. The theorem holds for $k = 0$. Assume now that, for all $0 \leqq n \leqq k$, every n-argument mapping $g\colon \boldsymbol{P}_m^n \to \boldsymbol{P}_m$ can be presented by means of the operations in $\boldsymbol{P}_m$. Let $f\colon \boldsymbol{P}_m^{k+1} \to \boldsymbol{P}_m$. We set $J_0(x) = \neg D_1(x)$, $J_i(x) = D_i(x) \cap \neg D_{i+1}(x)$, $i = 1, \cdots, m-2$, $J_{m-1}(x) = D_{m-1}(x)$ for each $x \in \boldsymbol{P}_m$. It can easily be verified that

$$J_i(e_j) = \begin{cases} e_{m-1} & \text{if} \quad i = j \\ e_0 & \text{if} \quad i \neq j \end{cases} \qquad i, j = 0, 1, \cdots, m-1.$$

It follows therefrom that $f(x_1, \cdots, x_k, x_{k+1}) = (f(x_1, \cdots, x_k, e_0) \cap$

$J_0(x_{k+1})) \cup \cdots \cup (f(x_1, \cdots, x_k, e_{m-1}) \cap J_{m-1}(x_{k+1}))$, since, for every $j = 0, \cdots, m-1$, if $x_{k+1} = e_j$, then $f(x_1, \cdots, x_k, e_j) \cap J_j(e_j) = f(x_1, \cdots, x_k, x_{k+1})$ is the only component different from e_0 on the right side of the equation given above.

The algebra $\boldsymbol{P}_m$ will be called the m-element Post algebra of order m. For $m = 2$, $\boldsymbol{P}_m$ is the two-element Boolean algebra.

The m-valued propositional calculi to be discussed here were constructed by Rousseau [45], [46]. Formalized languages $\mathcal{L}_m$ of these m-valued propositional calculi differ, for $m > 2$, from the language $\mathcal{L}_\chi$ of the intuitionistic propositional calculus by the fact that in addition to the logical operations $\cup, \cap, \Rightarrow, \neg$ they also include logical operations $D_1, \cdots, D_{m-1}$, of one argument each, and propositional constants $e_0, \cdots, e_{m-1}$. Every formula α of $\mathcal{L}_m$ will be interpreted algebraically in $\boldsymbol{P}_m$ in a way similar to that in which the formulas of $\mathcal{L}_\kappa$ are interpreted in the two-element Boolean algebra. Those formulas in $\mathcal{L}_m$ whose algebraic interpretations in $\boldsymbol{P}_m$ identically equal $e_{m-1} = \vee$ are laws (tautologies) of the m-valued propositional calculus under consideration.

The m-valued propositional calculi as defined above are functionally complete, which means that the abstract algebras $\boldsymbol{P}_m$ which determine these calculi are functionally complete. It is worth mentioning that they are equivalent with the functionally complete m-valued Łukasiewicz's propositional calculi with one designated value (investigated, e.g., by Rosser and Turquette [44]) in the following sense: all logical operations which occur in $\mathcal{L}_m$ are definable in Łukasiewicz's m-valued propositional calculus and conversely. All m-valued propositional calculi with logical values in $\boldsymbol{P}_m$, one of which $e_{m-1} = \vee$ is designated, are contained in the m-valued propositional calculi under consideration because any m-valued logical operation is definable in a functionally complete m-valued propositional calculus.

Rousseau's m-valued propositional calculi can be presented as formalized axiomatic systems. We adopt the axiom schemes (A_1)–(A_{10}), as in the case of the intuitionistic propositional calculus $\mathcal{S}_\chi$ (see Sec. 1), and additionally:

(A_{12}) $(D_i(\alpha \cup \beta) \Leftrightarrow (D_i(\alpha) \cup D_i(\beta)))$,
(A_{13}) $(D_i(\alpha \cap \beta) \Leftrightarrow (D_i(\alpha) \cap D_i(\beta)))$,

(A_{14}) $(D_i(\alpha \Rightarrow \beta) \Leftrightarrow (\cdots((D_1(\alpha) \Rightarrow D_1(\beta)) \cap \cdots \cap (D_i(\alpha) \Rightarrow D_i(\beta))))),$
(A_{15}) $(D_i(\neg\alpha) \Leftrightarrow \neg D_1(\alpha)),$
(A_{16}) $(D_i(D_j(\alpha)) \Leftrightarrow D_j(\alpha)),$
(A_{17}) $D_i(e_j)$ for $i \leqq j,$
(A_{18}) $\neg D_i(e_j)$ for $i > j,$
(A_{19}) $(\alpha \Leftrightarrow (\cdots(D_1(\alpha) \cap e_1) \cup \cdots \cup (\cdots(D_{m-1}(\alpha) \cap e_{m-1}))),$
(A_{20}) $(D_1(\alpha) \cup \neg D_1(\alpha)),$

where α, β, γ are any formulas in $\mathfrak{L}_m$, $i, j = 1, \cdots, m-1$ (except (A_{18}) where $j = 0, \cdots, m-1$)), and $\gamma \Leftrightarrow \delta$ is an abbreviation of $(\gamma \Rightarrow \delta) \cap (\delta \Rightarrow \gamma)$ for all formulas γ, δ in $\mathfrak{L}_m$.

In addition to the inference rule of modus ponens we adopt a rule (r_m), which to every formula α assigns the formula $D_{m-1}(\alpha)$. The formalized axiomatic system based on the above system of axioms and rules of inference will be denoted by $\mathcal{S}_m$ and called an m-valued propositional calculus. The following completeness theorem then holds: The set of all logical laws in $\mathfrak{L}_m$ coincides with the set of all theorems in $\mathcal{S}_m$ (Rousseau [45]). The proof refers to the idea used in the proof of Theorem 1 and to the concept of a Post algebra of order m.

The concept of Post algebras of order $m \geqq 2$ was first introduced by Rosenbloom [42], and later simplified successively by Epstein [10] and Traczyk [53], [55], [57]; the last-named also proved the equational definability of classes of Post algebras of any order $m \geqq 2$ and was the first to formulate systems of equational axioms. The theory of these algebras has been later developed in works of Traczyk ([53]–[60]), Chang and Horn [4], Dwinger [7]–[9], Malcev [24], Rousseau [45], [46], Sawicka [47], Włodarska [61], and others. A far-reaching generalization of this concept has been introduced by Traczyk [59], [60], Dwinger [8], [9], and recently by Cat Ho [3].

The concept of Post algebras of any order $m \geqq 2$ refers to that of a distributive lattice with the unit element $\vee$ and the zero element $\wedge$. Any such distributive lattice is an abstract algebra $(\boldsymbol{A}, \cup, \cap, \vee, \wedge)$ in which the operations of join ($\cup$) and of meet ($\cap$) follow laws analogous to those governing the set-theoretical operations of union and intersection, respectively. The class of all such distributive lattices is equationally definable[9]. Examples can be seen in rings of sets, i.e., non-empty families $\boldsymbol{R}(X)$ of subsets of any space X, including the empty set 0 and the whole space X, and such that the union and the intersection of any two sets in $\boldsymbol{R}(X)$ are also in $\boldsymbol{R}(X)$. By the representation theorem (Birkhoff [2], Stone [51]) for those distributive lattices the rings of sets are the only examples up to isomorphism. Any Boolean algebra is a distributive lattice with $\vee$ and $\wedge$. Note that in any distributive lattice a partial order relation $\leq$ (lattice ordering relation) can be defined by the equation (1).

Rousseau's system of axioms for Post algebras of order m is obtained when the equations $D_i(e_j) = e_{m-1} = \vee$ for $i \leqq j$, $D_i(e_j) = e_0 = \neg\vee$ for $i > j$, $D_1(a) \cup \neg D_1(a) = \vee$, and the system of equations obtained from schemata (A_{12})–(A_{16}), (A_{19}) by the replacement of the symbol $\Leftrightarrow$ by the identity symbol $(=)$ and of the symbols α, β, γ by a, b, c respectively, are joined to the axioms of pseudo-Boolean algebras (cf. Rasiowa and Sikorski [41]).

The m-element Post algebra $\boldsymbol{P}_m$ is the simplest example of a Post algebra of order m. It plays the same role in the theory of those algebras as does the two-element Boolean algebra in the theory of Boolean algebras. Note that any two m-element Post algebras of order m are isomorphic.

Every Post algebra

$$\boldsymbol{P} = (\boldsymbol{P}, \vee, \cup, \cap, \Rightarrow, \neg, D_1, \cdots, D_{m-1}, e_0, \cdots, e_{m-1})$$

is a pseudo-Boolean algebra with respect to the operations $\vee, \cup, \cap, \Rightarrow, \neg$. This implies (cf. Rasiowa and Sikorski [41]) that $(\boldsymbol{P}, \cup, \cap, \vee, \wedge)$, where $\wedge = \neg\vee$, is a distributive lattice with $\vee$ and $\wedge$. It is said that an element a in a distributive lattice with $\vee$ and $\wedge$ is complemented if there is an element b of that lattice such that $a \cup b = \vee$ and $a \cap b = \wedge$; in that case there is exactly one element b which has this property and it is called the complement of a. The complement of a complemented element a is denoted by $-a$. It is known that the set of all complemented

[9] A system of axioms is to be found, e.g., in [41].

elements of a distributive lattice with $\vee$ and $\wedge$ is a Boolean algebra. It turns out that the set $\boldsymbol{B_P}$ of all elements $D_i(a)$, where $a \in \boldsymbol{P}$ and $i = 1, \cdots, m-1$ for a Post algebra $\boldsymbol{P}$, coincides with the set of all complemented elements in the distributive lattice $(\boldsymbol{P}, \cup, \cap, \vee, \wedge)$ with $\vee$ and $\wedge$, and hence $(\boldsymbol{B_P}, \cup, \cap, \Rightarrow, \urcorner)$ is a Boolean algebra called the Boolean algebra corresponding to $\boldsymbol{P}$. It can be proved that every element $a \in \boldsymbol{P}$ has exactly one representation in the form: $a = (D_1(a) \cap e_1) \cup \cdots \cup (D_{m-1}(a) \cap e_{m-1})$, where $D_{m-1}(a) \leq \cdots \leq D_1(a)$ and $D_i(a) \in \boldsymbol{B_P}, i = 1, \cdots, m-1$. Note also that

$$(7) \qquad D_{m-1}(a) \leq a \leq D_1(a).$$

Thus every Post algebra $\boldsymbol{P}$ of order m uniquely determines a Boolean algebra $\boldsymbol{B_P} = (\boldsymbol{B_P}, \cup, \cap, \Rightarrow, \urcorner)$ and a chain $\wedge = e_0 \leq \cdots \leq e_{m-1} = \vee$. It can be shown, conversely, that this Boolean algebra, together with the chain considered above, determines a Post algebra of order m isomorphic with $\boldsymbol{P}$.

In order to explain the last statement observe that both the chain $\wedge = e_0 \leq \cdots \leq e_{m-1} = \vee$ and the Boolean algebra $\boldsymbol{B_P}$ are distributive lattices with $\vee$ and $\wedge$. The term co-product of two distributive lattices with $\vee$ and $\wedge$, $\boldsymbol{L_1} = (\boldsymbol{L_1}, \cup, \cap, \vee, \wedge)$ and $\boldsymbol{L_2} = (\boldsymbol{L_2}, \cup, \cap, \vee, \wedge)$, is used with reference to a distributive lattice $\boldsymbol{A} = (\boldsymbol{A}, \cup, \cap, \vee, \wedge)$ if there are two subalgebras of $\boldsymbol{A}$, $\boldsymbol{A_1} = (\boldsymbol{A_1}, \cup, \cap, \vee, \wedge)$ and $\boldsymbol{A_2} = (\boldsymbol{A_2}, \cup, \cap, \vee, \wedge)$ such that: (i) the union $\boldsymbol{A_1} \cup \boldsymbol{A_2}$ generates $\boldsymbol{A}$; (ii) $\boldsymbol{A_i}$ is isomorphic with $\boldsymbol{L_i}$, $i = 1, 2$; (iii) if h_i, $i = 1, 2$, is a homomorphism of $\boldsymbol{A_i}$ into a distributive lattice $\boldsymbol{C} = (\boldsymbol{C}, \cup, \cap, \vee, \wedge)$, then there is a homomorphism h of $\boldsymbol{A}$ into $\boldsymbol{C}$ which is a common extension of h_1 and h_2. It is known from the universal algebra that a co-product of any two distributive lattices with $\vee$ and $\wedge$ does exist and is determined up to isomorphism (Sikorski [48]). It can be proved that $\boldsymbol{P}$ is a co-product of $\boldsymbol{B_P}$ and the chain $\wedge = e_0 \leq \cdots \leq e_{m-1} = \vee$. This yields another equivalent definition of a Post algebra of order m.

Given a Boolean algebra $\boldsymbol{B} = (\boldsymbol{B}, \cup, \cap, \Rightarrow, -)$ we can construct a Post algebra $\boldsymbol{P} = (\boldsymbol{P}, \vee, \cup, \cap, \Rightarrow, \urcorner, D_1, \cdots, D_{m-1}, e_0, \cdots, e_{m-1})$ of order m, such that $\boldsymbol{B_P}$ is isomorphic with $\boldsymbol{B}$, using the following method. Let $\boldsymbol{P}$ be the set of all $m-1$ element sequences $(b_1, \cdots, b_{m-1})$, where for $1 \leqq i \leqq m-1$, $b_i \in \boldsymbol{B}$ and $b_{m-1} \leq \cdots \leq b_1$. Then $\boldsymbol{P}$ is a Post algebra of order m with respect to the operations defined thus: $\vee = (\vee, \cdots, \vee)$; $(b_1, \cdots, b_{m-1}) \cup (c_1, \cdots, c_{m-1}) = (b_1 \cup c_1, \cdots, b_{m-1} \cup c_{m-1})$; $(b_1, \cdots, b_{m-1}) \cap (c_1, \cdots, c_{m-1}) = (b_1 \cap c_1, \cdots, b_{m-1} \cap c_{m-1})$; $(b_1, \cdots, b_{m-1}) \Rightarrow (c_1, \cdots, c_{m-1}) = (b_1 \Rightarrow c_1, (b_1 \Rightarrow c_1) \cap (b_2 \Rightarrow c_2), \cdots, (b_1 \Rightarrow c_1) \cap \cdots \cap (b_{m-1} \Rightarrow c_{m-1}))$; $\neg(b_1, \cdots, b_{m-1}) = (\neg b_1, \cdots, \neg b_1)$; $D_i(b_1, \cdots, b_{m-1}) = (b_i, \cdots, b_i)$ for $i = 1, \cdots, m-1$; $e_0 = (\wedge, \cdots, \wedge)$; $e_1 = (\vee, \wedge, \cdots, \wedge)$; $e_2 = (\vee, \vee, \wedge, \cdots, \wedge)$, $\cdots, e_{m-1} = (\vee, \cdots, \vee)$. The Boolean algebra $\boldsymbol{B_P}$ consists of all elements $(b, \cdots, b)$ for $b \in \boldsymbol{B}$. Clearly $\boldsymbol{B_P}$ is isomorphic with $\boldsymbol{B}$.

Examples of Post algebras are also offered by Post fields of sets. They are constructed as follows. A topological space X is called a Post space of order $m \geqq 2$ if: (i) X is the set-theoretical union of disjoint topological spaces X_i, $1 \leqq i \leqq m-1$; (ii) there is a compact totally disconnected Hausdorff space X_0 and homeomorphisms $g_i: X_i \rightarrow X_0$ of X_i onto X_0, $1 \leqq i \leqq m-1$; (iii) the family $\boldsymbol{B}(X)$ of all subsets of X of the form $\bigcup_{i=1}^{m-1} g_i^{-1}(U)$, where U is both open and closed subset of X_0, is a basis for the open sets in X. It is proved that then $\boldsymbol{B}(X)$ is the field of all both open and closed subsets of X. We set $E_0 = 0$, $E_1 = X_1$, $E_2 = X_1 \cup X_2, \cdots, E_{m-1} = X_1 \cup \cdots \cup X_{m-1} = X$. Consider the family $\boldsymbol{P}(X)$ of all subsets Y of X of the form $Y = (Y_1 \cap E_1) \cup \cdots \cup (Y_{m-1} \cap E_{m-1})$, where $Y_i \in \boldsymbol{B}(X)$, $i = 1, \cdots, m-1$, and $Y_{m-1} \subset \cdots \subset Y_1$. In such a case $\boldsymbol{B}(X) \subset \boldsymbol{P}(X)$ and $E_i \in \boldsymbol{P}(X)$. It can be proved that for every $Y \in \boldsymbol{P}(X)$ there is exactly one representation of the form given above. We may thus introduce the operations D_i, $i = 1, \cdots, m-1$, by setting $D_i(Y) = Y_i$ for every $Y \in \boldsymbol{P}(X)$. Let us set, for any $Y, Z \in \boldsymbol{P}(X)$, $Y \Rightarrow Z = \bigcup_{i=1}^{m-1} (\bigcap_{j=1}^{i} (-D_j(Y) \cup D_j(Z)) \cap E_i)$ and $\neg Y = Y \Rightarrow E_0$. It can be proved that $(\boldsymbol{P}(X), X, \cup, \cap, \Rightarrow, \neg, D_1, \cdots, D_{m-1}, E_0, \cdots, E_{m-1})$ is a Post algebra of order m which is called a Post field of sets (or order m). The representation theorem for Post algebras of order m states that each Post algebra is isomorphic with a Post field of sets of the same order (Traczyk [55], Dwinger [7]).

A mapping h of a Post algebra $\boldsymbol{P}$ of order m onto a Post algebra

$\boldsymbol{P}'$ of the same order is called an epimorphism if $h(a \cup b) = h(a) \cup h(b)$, $h(a \cap b) = h(a) \cap h(b)$, $h(a \Rightarrow b) = h(a) \Rightarrow h(b)$, $h(\neg a) = \neg h(a)$, $h(D_i(a)) = D_i(h(a))$, $i = 1, \cdots, m-1$, $h(e_i) = e_i'$ in $\boldsymbol{P}'$ for $i = 0, \cdots, m-1$. In the case of Post algebras D-filters[10] play a role which is analogous to that of filters in Boolean algebras. D-filters are non-empty subsets ∇ of the set of elements of a Post algebra $\boldsymbol{P}$ (of order m) which are filters in the sense adopted for Boolean algebras and have the additional property that if $a \in \nabla$, then $D_{m-1}(a) \in \nabla$. Kernels of epimorphisms (i.e., inverse images of the unit element $\vee$) of Post algebras are D-filters and conversely. Every D-filter ∇ in $\boldsymbol{P}$ determines a congruence relation $\approx_\nabla$ defined as in Boolean algebras. The quotient algebra denoted by $\boldsymbol{P}/\nabla$ is a Post algebra of the same order. The mapping h defined by the equation $h(a) = [a] \in \boldsymbol{P}/\nabla$ for $a \in \boldsymbol{P}$ is an epimorphism from $\boldsymbol{P}$ onto $\boldsymbol{P}/\nabla$, whose kernel is ∇.

Prime D-filters (i.e. proper D-filters satisfying the condition (3), Sec. 1) coincide with kernels of epimorphisms of Post algebras of order m onto the m-element Post algebra $\boldsymbol{P}_m$ of the same order. There is a close connection between prime D-filters in a Post algebra $\boldsymbol{P}$ and prime filters in the corresponding Boolean algebra $\boldsymbol{B}_{\boldsymbol{P}}$. The following theorem holds:

THEOREM 4: *If ∇ is a prime D-filter in a Post algebra $\boldsymbol{P}$, then $\nabla_0 = \nabla \cap \boldsymbol{B}_{\boldsymbol{P}}$ is a prime filter in the corresponding Boolean algebra $\boldsymbol{B}_{\boldsymbol{P}}$. Conversely, if ∇_0 is a prime filter in $\boldsymbol{B}_{\boldsymbol{P}}$, then the set $\nabla = \{a \in \boldsymbol{P}\colon D_{m-1}(a) \in \nabla_0\}$ is a prime D-filter in $\boldsymbol{P}$ and $\nabla_0 = \nabla \cap \boldsymbol{B}_{\boldsymbol{P}}$.*

It follows easily from (4) and Theorem 4 that

(8) for each element $a \neq \vee$ in a Post algebra $\boldsymbol{P}$ of order m there is a prime D-filter ∇ such that $a \notin \nabla$.

In fact, if $a \neq \vee$, then by (7), $D_{m-1}(a) \neq \vee$. In consequence, $\neg D_{m-1}(a) \neq \wedge$ in the Boolean algebra $\boldsymbol{B}_{\boldsymbol{P}}$. On account of (4) there is a prime filter ∇_0 in $\boldsymbol{B}_{\boldsymbol{P}}$ such that $\neg D_{m-1}(a) \in \nabla_0$, so that

[10] The notion of Post epimorphism as adopted here (see [37]) differs from that used by Traczyk. D-filters (see [37]) are dual to Traczyk's ideals of order 1.

$D_{m-1}(a) \notin \nabla_0$. The prime D-filter $\nabla = \{b \in \boldsymbol{P}\colon D_{m-1}(b) \in \nabla_0\}$ does not contain a.

Now we revert to our analysis of the algebraic interpretation of the m-valued propositional calculi $\mathcal{S}_m$, $m \geqq 2$. Formulas in $\mathcal{S}_m$ may be interpreted in any Post algebra of order m, just as formulas in the two-valued propositional calculus $\mathcal{S}_\kappa$ may be interpreted in any Boolean algebra. The following (see [45]) is a generalized version of Theorem 1.

THEOREM 5: *For any formula α in the m-valued propositional calculus $\mathcal{S}_m$ the following conditions are equivalent:*

(i) $\vdash \alpha$,
(ii) $\alpha_{\boldsymbol{P}} = \vee$ *identically in every Post algebra of order m,*
(iii) $\alpha_{\boldsymbol{P}(X)} = X$ *identically in every Post field $\boldsymbol{P}(X)$ of order m,*
(iv) $\alpha_{\boldsymbol{P}_m} = \vee$ *identically in the m-element Post algebra $\boldsymbol{P}_m$.*

The proofs of the implications (i) $\rightarrow$ (ii), (ii) $\rightarrow$ (iii), (iii) $\rightarrow$ (iv) do not present any difficulties. The proof of (iv) $\rightarrow$ (i) is analogous to that in Theorem 1. An algebra $\boldsymbol{A}(\mathcal{S}_m)$ of $\mathcal{S}_m$ is constructed in the same way as the algebra $\boldsymbol{A}(\mathcal{S}_\kappa)$. Such an algebra is a Post algebra or order m. It is proved that $\| \alpha \| = \vee$ iff $\vdash \alpha$, for any formula α in $\mathcal{S}_m$. Assume that not $\vdash \alpha$. Hence $\| \alpha \| \neq \vee$. By (8) there is a prime D-filter ∇ such that $\| \alpha \| \notin \nabla$. The quotient algebra $\boldsymbol{A}(\mathcal{S}_m)/\nabla$ is isomorphic with $\boldsymbol{P}_m$. Moreover, since (2) holds for Post algebras, too, its element $[\| \alpha \|] \neq \vee$. For the valuation v_0 in $\boldsymbol{A}(\mathcal{S}_m)/\nabla$ defined by the equation $v_0(p_i) = [\| p_i \|]$, $i = 1, 2, \cdots$, we have, for every formula β, $\beta_{\boldsymbol{A}(\mathcal{S}_m)/\nabla}(v_0) = [\| \beta \|]$. Hence $\alpha_{\boldsymbol{A}(\mathcal{S}_m)/\nabla} = [\| \alpha \|] \neq \vee$.

An interesting conclusion follows from Theorem 5 and Theorem 3:

THEOREM 6: *For any formula α in the linear intuitionistic propositional calculus $\mathcal{S}_{\chi l}$ the following conditions are equivalent:*

(i) *α is a theorem in $\mathcal{S}_{\chi l}$,*
(ii) *for every $m = 2, 3, \cdots$, α is a theorem of $\mathcal{S}_m$,*
(iii) *α is a theorem in $\mathcal{S}_m$ for each $2 \leqq m \leqq n + 2$, where n is the number of variables in α.*

3. AN ALGEBRAIC TREATMENT OF ELEMENTARY FORMALIZED THEORIES BASED ON VARIOUS LOGICS

The question arises naturally as to how more comprehensive logical systems, namely first-order predicate calculi and elementary formalized theories, are to be treated from the algebraic point of view. The idea of generalizing the algebraic interpretation of formulas so as to cover predicate calculus formulas is due to A. Mostowski [50], who suggested such an interpretation for an intuitionistic predicate calculus $\mathcal{S}_\chi^*$ and posed the question whether that interpretation fully characterized the theorems of that system. Mostowski's idea will now be described briefly.

The predicate calculi differ from the propositional calculi of the same logic on the following points:

(1) The formulas are constructed not of propositional variables, but of what is called atomic formulas in the form $\rho_n(x_{k_1}, \cdots, x_{k_{m_n}})$, where ρ_n is a predicate (relation symbol) of m_n arguments $(n = 1, 2, \cdots; m_n = 1, 2, \cdots)$, and $x_1, x_2, \cdots$ are free individual variables.

(2) In addition to the logical operations that occur in a given propositional calculus, the symbols $\bigcup\xi_k$, $\bigcap\xi_k$, $k = 1, 2, \cdots$, which correspond, respectively, to the expressions "there is an ξ_k", "for every ξ_k", are introduced ($\xi_1, \xi_2, \xi_3, \cdots$ being the bound individual variables); they are called the existential and the universal quantifiers binding ξ_k, respectively; if $\alpha(x_i)$ is a formula in which a free individual variable x_i occurs and neither $\bigcup\xi_k$ nor $\bigcap\xi_k$ does, then $\bigcup\xi_k\alpha(\xi_k)$ and $\bigcap\xi_k\alpha(\xi_k)$ are formulas.

(3) Logical axiom-schemes are obtained from those adopted in a given propositional calculus by treating $\alpha, \beta, \gamma, \cdots$ which occur in them as arbitrary predicate calculus formulas.

(4) The list of the rules of inference adopted in a given propositional calculus is augmented by rules on the quantifiers and the rule on substitution for free individual variables as in the classical predicate calculus (cf. Rasiowa and Sikorski [41]).

Those classes of algebras in which the propositional calculi mentioned in Sec. 1 have an interpretation have the following property: the algebras which are in those classes, or, more strictly speaking, their respective reducts, are distributive lattices. Hence in each of them we can define a partial ordering (lattice ordering) $\leq$ and consider the concepts of least upper bound and greatest lower bound. If, for every set $\{a_u\}_{u\in U}$ of elements in a given algebra $\boldsymbol{A}$ there is a least upper bound denoted by $\bigcup^{\boldsymbol{A}}_{u\in U} a_u$ (called also an infinite join) and a greatest lower bound denoted by $\bigcap^{\boldsymbol{A}}_{u\in U} a_u$ [11] (called also an infinite meet), then such an algebra is called complete. In particular, we may speak about complete Boolean algebras, pseudo-Boolean algebras, topological Boolean algebras, relatively pseudo-complemented lattices, μ-algebras, N-lattices, Post algebras.

Mostowski's idea of an algebraic interpretation of the formulas of the intuitionistic predicate calculus $\mathcal{S}^*_\chi$ (obtained from $\mathcal{S}_\chi$ by the method described above) is based on the following principle. A non-empty set J (universe) and a complete pseudo-Boolean algebra $\boldsymbol{A}$ is fixed. Free individual variables are interpreted as variables which range over J, the predicates ρ_n as m_n-argument functions from J into $\boldsymbol{A}$, the logical propositional operations as the corresponding algebraic operations in $\boldsymbol{A}$, and the quantifiers $\bigcup \xi_k$, $\bigcap \xi_k$ as infinite joins and meets in $\boldsymbol{A}$, respectively. The assumption that $\boldsymbol{A}$ is complete guarantees the existence of all infinite joins and meets which occur in the interpretation of the formulas.

Mostowski has proved that if α is a theorem in $\mathcal{S}^*_\chi$, then every algebraic interpretation $\alpha_\mathfrak{s}$ of that formula in any set J and in any complete pseudo-Boolean algebra $\boldsymbol{A}$ identically equals $\vee$. The converse theorem, which answers the question posed by A. Mostowski, has been proved by the present author [33]. In that paper an analogous interpretation of the formulas of a modal predicate calculus $\mathcal{S}^*_\lambda$ [12] in complete topological Boolean algebras

[11] The symbol $\boldsymbol{A}$ of the algebra under consideration will be omitted if no ambiguity results from this.

[12] This predicate calculus corresponds to the Lewis modal propositional calculus S4 (Lewis and Langford [17]).

has been used, and the theorem stating that any formula in $\mathcal{S}_\chi^*$ is a theorem iff any of its algebraic interpretations in any non-empty set J and in any complete topological Boolean algebra identically equals $\vee$, has been proved. The proof of the theorem converse to that of Mostowski refers to the idea described in the proof of Theorem 1. An algebra $\boldsymbol{A}(\mathcal{S}_\chi^*)$ of the system $\mathcal{S}_\chi^*$ is constructed as for $\mathcal{S}_\kappa$, and it is then proved that it is a pseudo-Boolean algebra. This algebra is not complete, but it has the following infinite joins and meets that correspond to the quantifiers which occur in any formula:

$$(\mathrm{Q}) \qquad \begin{aligned} \bigcup_{n \in N} \| \gamma(x_n) \| &= \| \bigcup \xi_k \gamma(\xi_k) \|, \\ \bigcap_{n \in N} \| \gamma(x_n) \| &= \| \bigcap \xi_k \gamma(\xi_k) \|^{13}, \end{aligned}$$

where N denotes the set of all positive integers and γ an arbitrary formula in $\mathcal{S}_\chi^*$. It is proved that every pseudo-Boolean algebra, in particular $\boldsymbol{A}(\mathcal{S}_\chi^*)$, can be extended into a complete pseudo-Boolean algebra so that all infinite joins and meets in a given algebra remain unchanged under that extension. Such an extension $\boldsymbol{A}$ of $\boldsymbol{A}(\mathcal{S}_\chi^*)$ is adopted as a complete algebra which has to be fixed in order to define an interpretation of formulas. As a set J we adopt the set $\{x_n\}_{n \in N}$ of all free individual variables. Then the following interpretation $\mathcal{J}$ in J and $\boldsymbol{A}$ is called canonical for $\mathcal{S}_\chi^*$: for every predicate ρ_n, its interpretation $\rho_{n,\mathcal{J}}(x_{k_1}, \cdots, x_{k_{mn}}) = \| \rho_n(x_{k_1}, \cdots, x_{k_{mn}}) \| \in \boldsymbol{A}$. This interpretation has the following property: if with every variable x_n, $n \in N$, is assigned a variable $s(x_n)$ as its value in J, then, for every formula β in $\mathcal{S}_\chi^*$, its canonical interpretation $\beta_{\mathcal{J}}(s) = \| s\beta \|$, where $s\beta$ denotes a formula obtained from β by the replacement of each occurrence of x_n $(n \in N)$ in β by $s(x_n)$. In particular, if $s_0(x_n) = x_n$ for every $n \in N$, then $\beta_{\mathcal{J}}(x_0) = \| \beta \|$. In the proof of this property of a canonical interpretation, an essential role is played by the equations (Q) and the fact that they are preserved when the algebra $\boldsymbol{A}(\mathcal{S}_\chi^*)$ is extended into a complete algebra $\boldsymbol{A}$.

[13] Analogous equations were obtained by Henkin [11], who examined a predicate calculus weaker than $\mathcal{S}_\chi^*$, namely that of positive implicative logic.

If not $\vdash\alpha$, then $\| \alpha \| \neq \vee$, from which it follows that $\alpha_{\mathcal{J}}(s_0) \neq \vee$, so that there is an interpretation $\mathcal{J}$ of the formulas in a complete pseudo-Boolean algebra $\boldsymbol{A}$ and in a set J such that $\alpha_{\mathcal{J}}$ does not identically equal $\vee$. The proof of the analogous statement for the formulas of $\mathcal{S}_{\chi}^{*}$ is based on the same idea. The proof of the converse statement is very simple.

The method outlined above has led to many papers on the applications of abstract algebra to mathematical logic. The results obtained so far make it possible to treat a large part of mathematical logic in a uniform manner, namely to reduce logical problems to those of lattice theory. A systematic exposition is to be found in *The Mathematics of Metamathematics* by the present author and R. Sikorski.

Only some results thus obtained will be given here in a cursory review in order to introduce the reader to problems to be considered in the next section.

In 1950, Rasiowa and Sikorski found the first algebraic proof of the Gödel completeness theorem for the first-order classical predicate calculus $\mathcal{S}_{\kappa}^{*}$. The intuitive content of this theorem is that a formula α in $\mathcal{S}_{\kappa}^{*}$ is a theorem iff, by any interpretation of the relation symbols ρ_n which occur in α as m_n-argument relations in the set N of positive integers, α presents a propositional function true in N. More precisely this theorem states that a necessary and sufficient condition for any formula α to be a theorem in $\mathcal{S}_{\kappa}^{*}$ is that any of its algebraic interpretations in the two-element Boolean algebra $\boldsymbol{B}_0$ (which is complete) and in the set N of positive integers identically equals $\vee$. The proof uses the idea described in the proof of Theorem 1 together with the method indicated in our comments on the intuitionistic predicate calculus $\mathcal{S}_{\chi}^{*}$. An algebra $\boldsymbol{A}(\mathcal{S}_{\kappa}^{*})$ is constructed for the system $\mathcal{S}_{\kappa}^{*}$ (as for $\mathcal{S}_{\kappa}$) which is a Boolean algebra and in which the infinite joins and meets (Q) exist. If α is not a theorem in $\mathcal{S}_{\kappa}^{*}$, then $\| \alpha \| \neq \vee$ so that $-\| \alpha \| \neq \wedge$. By (4) in Sec. 1 there is in that algebra a prime filter ∇_0 such that $-\| \alpha \| \in \nabla_0$ which implies that $\| \alpha \| \notin \nabla_0$. Unfortunately such a prime filter need not have the property that the equations (Q) are preserved in the quotient algebra $\boldsymbol{A}(\mathcal{S}_{\kappa}^{*})/\nabla_0$, i.e., that the equations

$$(Q^*)\qquad \begin{aligned} \bigcup_{n\in N}[\| \gamma(x_n) \|] &= [\| \bigcup_{\xi_k}\gamma(\xi_k) \|], \\ \bigcap_{n\in N}[\| \gamma(x_n) \|] &= [\| \bigcap_{\xi_k}\gamma(\xi_k) \|], \end{aligned}$$

hold in $A(S_\kappa^*)/\nabla_0$. The following lemma (Rasiowa and Sikorski, 1950[14]) on Boolean algebras makes it possible to eliminate this difficulty. If, in a Boolean algebra A, there is given an element $a \neq \wedge$ (an element $b \neq \vee$) and a denumerable set (S) of infinite joins and infinite meets, then there is a prime filter ∇ in A such that $a \in \nabla$ ($b \notin \nabla$) and all those infinite joins and infinite meets which occur in (S) are preserved in A/∇.

We accordingly consider a prime filter ∇ such that $\| \alpha \| \notin \nabla$ and such that the equations (Q*) hold in $A(S_\kappa^*)/\nabla$, which is the two-element Boolean algebra. An interpretation $\mathcal{J}$ in $A(S_\kappa^*)/\nabla$ and in the set $J = \{x_n)_{n\in N}$ that corresponds to a canonical one, i.e., such that, for every predicate ρ_n, $\rho_{n,\mathcal{J}}(x_{k_1}, \cdots, x_{k_{m_n}}) = [\| \rho_n(x_{k_1}, \cdots, x_{k_{m_n}}) \|]$ has the following property: for every formula β in S_κ^* and for the identity mapping s_0 from the set J into J we have $\beta_{\mathcal{J}}(s_0) = [\| \beta \|]$. In the proof of this statement, the equations (Q*) play an essential part. In particular, $\alpha_{\mathcal{J}}(s_0) = [\| \alpha \|] \neq \vee$ because $\| \alpha \| \notin \nabla$ (cf. (2) in Sec. 1). Hence, if α is not a theorem in S_κ^*, then there is an enumerable set $J = \{x_n\}_{n\in N}$ and an interpretation $\mathcal{J}$ in the two-element Boolean algebra $A(S_\kappa^*)/\nabla$ and J such that $\alpha_{\mathcal{J}}$ does not identically equal $\vee$. A transition from J to the set N of positive integers presents no difficulty; nor does the proof of the converse implication. Other algebraic proofs of the Gödel theorem have been later published by various authors.

In 1951, Rasiowa and Sikorski gave a simple algebraic proof of the Löwenheim-Skolem-Gödel theorem for the classical predicate calculi by applying a method similar to that used by them in the proof of the Gödel completeness theorem. That theorem states that

---

[14] Rasiowa and Sikorski, Fundamenta Mathematicae, Vol. 37 (1950), pp. 193–200.

every consistent set of closed formulas (i.e., formulas without free individual variables) is simultaneously satisfiable in the set of positive integers, which means that there is an algebraic interpretation $\mathcal{I}$ of the predicates in the two-element Boolean algebra $\boldsymbol{B}_0$ and the set N of positive integers such that, for every formula α in the set of formulas under consideration, $\alpha_{\mathcal{I}} = \vee$. Other algebraic proofs of this theorem have been published later. Mention is due here to Łoś's proof [18] which consists in a transition to the case in which only open formulas (i.e., those without quantifiers) are considered. This transition can be achieved by reference to ϵ-theorems (Hilbert and Bernays [14]). The proof for the case of open formulas is quite similar to that of Theorem 1.

The Gödel completeness theorem and the above-mentioned results obtained concerning the intuitionistic predicate calculus $\mathcal{S}^*_\chi$ were generalized by Rasiowa and Sikorski in 1953, who proved the following theorem:

THEOREM 7: *For every formula α in $\mathcal{S}^*_\kappa$ (in $\mathcal{S}^*_\chi$) the following conditions are equivalent:*

(i) *α is a theorem in $\mathcal{S}^*_\kappa$ (in $\mathcal{S}^*_\chi$),*

(ii) *$\alpha_{\mathcal{I}} = \vee$ identically for every interpretation $\mathcal{I}$ in every set $J \neq 0$ and in every complete Boolean (pseudo-Boolean) algebra $\boldsymbol{A}$,*

(iii) *$\alpha_{\mathcal{I}} = X$ identically for every interpretation $\mathcal{I}$ in every set $J \neq 0$ and in every field (pseudo-field) of all subsets (open subsets) of any space X (any topological space X),*

(iv) *$\alpha_{\mathcal{I}} = X$ identically for every interpretation $\mathcal{I}$ in the set N of positive integers and in every field (pseudo-field) of all subsets (open subsets) of any space X (any topological space X).*

The following lemma on Boolean algebras (pseudo-Boolean algebras) due to these authors is a basis of the proof of Theorem 7.[15] If, in a Boolean (pseudo-Boolean) algebra $\boldsymbol{A}$, there is given an enumerable set (S) of infinite joins and meets, then there is a

[15] In the case of $\mathcal{S}^*_\kappa$ the lemma makes it possible to prove the equivalence of conditions (i) and (iv) independently of the Gödel completeness theorem and to obtain a modified algebraic proof of the theorem (cf. [40] and [41]).

one-to-one homomorphism h of $\boldsymbol{A}$ into a field (pseudo-field) of all subsets (open subsets) of a space X (topological space X) which preserves all those infinite joins and meets which occur in (S). This means that the infinite joins and meets in (S) are mapped by h onto the corresponding set-theoretical unions and intersections (unions and interiors of intersections), respectively.

They also proved that the Löwenheim-Skolem-Gödel theorem can be generalized so as to cover the intuitionistic predicate calculus $\mathcal{S}_\chi^*$. Analogous results were obtained by these authors for the predicate calculi $\mathcal{S}_\mu^*$ of minimal logic, $\mathcal{S}_\pi^*$ of positive logic, $\mathcal{S}_\lambda^*$ of modal logic (see Rasiowa and Sikorski [40]). Sikorski [49] obtained a stronger result by demonstrating that in condition (iv) of Theorem 7, which concerns a formula α in $\mathcal{S}_\chi^*$, a set of irrational numbers can be taken as topological space X. The same applies to the calculi $\mathcal{S}_\mu^*$, $\mathcal{S}_\pi^*$ and $\mathcal{S}_\lambda^*$. The problem whether the whole straight line can be taken as X remains open for $\mathcal{S}_\chi^*$, $\mathcal{S}_\mu^*$, $\mathcal{S}_\pi^*$. For $\mathcal{S}_\lambda^*$ the answer is in the negative (see Rasiowa and Sikorski [40]).

A predicate calculus $\mathcal{S}^*$ of constructive logic with strong negation was investigated from the algebraic point of view by Białynicki-Birula and Rasiowa [1].

By singling out, in any predicate calculus $\mathcal{S}^*$, a set $\mathcal{A}$ of formulas which are called specific axioms we obtain a formalized elementary theory $\mathcal{S}^*(\mathcal{A})$, which has as its theorems all the formulas derivable from logical axioms and from specific axioms under the adopted rules of inference. Languages of elementary formalized theories are usually made richer by including in addition to predicates, symbols ϕ_n of functions of k_n arguments $(n = 1, 2, \cdots; k_n = 0, 1, 2, \cdots)$ called k_n-argument functors. In such a case, functors and free individual variables are used to construct naturally expressions called terms. Thus, e.g., if ϕ_1 is a functor of two arguments, and ϕ_2, ϕ_3 are functors of one argument each, then $\phi_1(\phi_2(x_1), \phi_3(x_2))$ is a term. Atomic formulas are then in the form $\rho_n(\tau_1, \cdots, \tau_{m_n})$, where $\tau_1, \cdots, \tau_{m_n}$ are any terms, and the remaining formulas are constructed of atomic ones in the same way as in the case of functor-free languages.

Functors of k_n-arguments each are interpreted, for a fixed set $J \neq 0$ (the universe), as functions of k_n arguments each from J

into J. The algebraic interpretation of functor-free formulas is similarly extended so as to cover formulas with functors occurring in them.

The concept of semantic model is a fundamental one in the methodology of the classical elementary formalized theories. A semantic model of a theory $\mathcal{S}^*_\kappa(\mathfrak{A})$ is an interpretation $\mathcal{J}$, in any set $J \neq 0$ and in the two-element Boolean algebra $\boldsymbol{B}_0$, such that, for every formula α in $\mathfrak{A}$, $\alpha_{\mathcal{J}} = \vee$ identically.

In 1955, the concept of algebraic model of any elementary formalized theory, which is a generalized version of that of semantic model, was introduced (Rasiowa [34]). An algebraic model of a theory $\mathcal{S}^*(\mathfrak{A})$ is any algebraic interpretation $\mathcal{J}$ of predicates and functors, in any set $J \neq 0$ and in a complete algebra $\boldsymbol{A}$ (which is in the class of those algebras in which the propositional calculi of a logic under consideration are interpreted), such that for every formula $\alpha \in \mathfrak{A}$, $\alpha_{\mathcal{J}} = \vee$ identically.

Algebraic models have properties that are analogous to those of semantic models. For instance, every consistent theory, i.e., every theory in which not every formula is a theorem, has an algebraic model. In the case of classical theories, this concept of consistency coincides with the concept of consistency in the sense that, for every formula α, at least one of the formulas α and $-\alpha$ is not a theorem. The fact that a theory has an algebraic model guarantees the consistency of that theory if the theory contains the logical operation of negation or seminegation, the latter being a unary logical operation with the property that the semi-negation of a true proposition is a false proposition. This holds in all logical systems mentioned in this paper, except for positive logic and minimal logic. In the case of classical theories, the existence of an algebraic model is equivalent to the existence of a semantic model. It is worth mentioning that every consistent theory has an adequate algebraic model, i.e., a model $\mathcal{J}$ such that, for every formula α, α is a theorem in the theory in question iff $\alpha_{\mathcal{J}} = \vee$ identically. Note that classical elementary theories mostly do not have adequate semantic models. Adequate algebraic models are a convenient tool in the proofs of various theorems in the methodology of elementary theories; in particular they have been used in the proofs of many theorems pertaining to the methodology of non-classical elementary theories.

The theory of Boolean models (i.e., of algebraic models of classical theories) has been much developed in recent years and used in simplified independence proofs in set-theory, in particular in a proof of the independence of the continuum hypothesis by D. Scott (cf. Rosser [43]).

4. AN ALGEBRAIC APPROACH TO ELEMENTARY FORMALIZED THEORIES BASED ON m-VALUED LOGICS

We now proceed to discuss applications of the theory of Post algebras of order m to the methodology of formalized elementary theories based on m-valued logics (cf. Sec. 2). To ensure complete generality of our analysis we extend the languages of the theories under consideration by admitting non-enumerable sets of individual variables, predicates, and functors. In accordance with the above we assume a formalized language $\mathfrak{L}_m$ which is to have the following properties. The sets V of all free individual variables and Ξ of all bound individual variables are infinite. The sets Φ_n of n-argument functors ($n = 0, 1, 2, \cdots$) in $\mathfrak{L}_m$ are of any powers; in particular, they can be empty. The sets P_n of n-argument predicates ($n = 1, 2, \cdots$) in $\mathfrak{L}_m$ are of any powers, but at least one of them is non-empty. The set T of all terms in $\mathfrak{L}_m$ includes V and Φ_0 and satisfies the condition: if $\phi \in \Phi_n$ and $\tau_1, \cdots, \tau_n \in T$, then $\phi(\tau_1, \cdots, \tau_n) \in T$. The set F_m of all formulas in $\mathfrak{L}_m$ contains all atomic formulas, i.e., formulas in the form $\rho(\tau_1, \cdots, \tau_n)$, where $\rho \in P_n$, $\tau_1, \cdots, \tau_n \in T$, $n = 1, 2, \cdots$, and formulas in the form e_i, $i = 0, \cdots, m-1$, and satisfies the following conditions: (i) if $\alpha, \beta \in F_m$, then $\neg\alpha$, $D_i(\alpha)$, $(i = 1, \cdots, m-1)$, $(\alpha \cup \beta)$, $(\alpha \cap \beta)$, $(\alpha \Rightarrow \beta) \in F_m$; (ii) if $\alpha(x) \in F_m$, $x \in V$, $\xi \in \Xi$, and $\bigcup\xi$, $\bigcap\xi$ do not occur in $\alpha(x)$, then $\bigcup\xi\alpha(\xi)$ and $\bigcap\xi\alpha(\xi)$ are in F_m. For brevity $(\alpha \Leftrightarrow \beta)$ will be used for $((\alpha \Rightarrow \beta) \cap (\beta \Rightarrow \alpha))$.

By making use of $\mathfrak{L}_m$ we construct an m-valued predicate calculus $\mathcal{S}_m^*$ on the basis of $\mathcal{S}_m$ (see Sec. 2) in the same manner as $\mathcal{S}_\kappa^*$ was constructed on the basis of $\mathcal{S}_\kappa$ (see Sec. 3). Thus as logical axioms we take all formulas in the form of (A_1)–(A_{10}) in Sec. 1 and (A_{12})–(A_{20}) in Sec. 2, where $\alpha, \beta, \gamma \in F_m$, and the rules of inference we take to be the same as in the propositional calculus $\mathcal{S}_m$ plus the rule of substitution for free individual variables of any terms and the rules of handling the quantifiers (cf. Rasiowa and Sikorski [41]). Every set $\mathfrak{A} \subset F_m$ determines an elementary theory $\mathcal{S}_m^*(\mathfrak{A})$ with the set $\mathfrak{A}$ as specific axioms. We write $\mathfrak{A} \vdash \alpha$ if α is a theorem in $\mathcal{S}_m^*(\mathfrak{A})$; in particular, we write $\vdash \alpha$ if α is a theorem in $\mathcal{S}_m^*(O) = \mathcal{S}_m^*$.

We introduce a relation $\approx_{\mathfrak{A}}$ in the set F_m, which holds between formulas α, β if $\mathfrak{A} \vdash (\alpha \Rightarrow \beta)$ and $\mathfrak{A} \vdash (\beta \Rightarrow \alpha)$. This relation is a congruence in the algebra of formulas $(F_m, \vee, \cup, \cap, \Rightarrow, \neg, D_1, \cdots, D_{m-1}, e_0, \cdots, e_{m-1})$, where $\vee = e_{m-1}$, and the quotient algebra denoted by $\boldsymbol{A}(\mathcal{S}_m^*(\mathfrak{A}))$ and called the algebra of the theory $\mathcal{S}_m^*(\mathfrak{A})$ is a Post algebra of order m. Moreover in this algebra

$$\text{(9)} \qquad \| \alpha \|_{\mathfrak{A}} = \vee \quad \text{iff} \quad \mathfrak{A} \vdash \alpha,$$

$$\text{(10)} \qquad \| \alpha \|_{\mathfrak{A}} \leq \| \beta \|_{\mathfrak{A}} \quad \text{iff} \quad \mathfrak{A} \vdash (\alpha \Rightarrow \beta).$$

It follows from (9) that $\boldsymbol{A}(\mathcal{S}_m^*(\mathfrak{A}))$ is non-degenerate (i.e., it has at least two elements) iff the theory $\mathcal{S}_m^*(\mathfrak{A})$ is consistent, which means that at least one formula is not a theorem in $\mathcal{S}_m^*(\mathfrak{A})$. The algebra under consideration is not complete, but it has the following infinite joins and meets that correspond to the quantifiers which occur in any formula:

$$(\mathrm{Q}_{\mathfrak{A}}) \qquad \begin{aligned} \bigcup_{\tau \in T} \| \gamma(\tau) \|_{\mathfrak{A}} &= \| \bigcup_{\xi} \gamma(\xi) \|_{\mathfrak{A}}, \\ \bigcap_{\tau \in T} \| \gamma(\tau) \|_{\mathfrak{A}} &= \| \bigcap_{\xi} \gamma(\xi) \|_{\mathfrak{A}}, \end{aligned}$$

where γ is an arbitrary formula in $\mathcal{S}_m^*(\mathfrak{A})$. The question arises whether every Post algebra of any order m (in particular $\boldsymbol{A}(\mathcal{S}_m^*(\mathfrak{A}))$) can be extended to a complete Post algebra of the same order with the preservation of all infinite joins and meets which exist in a given algebra. To answer this question we make use of the following

Epstein's Lemma [10]: *In every Post algebra* $\boldsymbol{P}$ *of order* m

(i) *if* $a = \bigcup_{u \in U}^{\boldsymbol{P}} a_u$, *then* $D_i(a) = \bigcup_{u \in U}^{\boldsymbol{B_P}} D_i(a_u)$, $i = 1, \cdots, m-1$,

(ii) *if* $b = \bigcap_{u \in U}^{\boldsymbol{P}} b_u$, *then* $D_i(b) = \bigcap_{u \in U}^{\boldsymbol{B_P}} D_i(b_u)$, $i = 1, \cdots, m-1$,

(iii) *if* a_u, $u \in U$, *are any elements in* $\boldsymbol{P}$, *and if for every* $i = 1, \cdots, m-1$, *there is in* $\boldsymbol{B_P}$, $\bigcup_{u \in U}^{\boldsymbol{B_P}} D_i(a_u) = a_i$ *then for the element* $a = (a_1 \cap e_1) \cup \cdots \cup (a_{m-1} \cap e_{m-1})$ *in* $\boldsymbol{P}$ *it is also true that* $a = \bigcup_{u \in U}^{\boldsymbol{P}} a_u$ *and* $a_i = D_i(a)$,

(iv) *if* b_u, $u \in U$, *are any elements in* $\boldsymbol{P}$ *and for every* $i = 1, \cdots, m-1$, *there is* $\bigcap_{u \in U}^{\boldsymbol{B_P}} D_i(b_u) = b_i$, *then for* $b = (b_1 \cap e_1) \cup \cdots \cup (b_{m-1} \cap e_{m-1})$ *the equation* $b = \bigcap_{u \in U}^{\boldsymbol{P}} b_u$ *holds and* $b_i = D_i(b)$.

It follows immediately from Epstein's lemma that a Post algebra $\boldsymbol{P}$ is complete iff its corresponding Boolean algebra $\boldsymbol{B_P}$ is complete.

Consider now any incomplete Post algebra $\boldsymbol{P}$ of order m and assume that $a = \bigcup_{u \in U}^{\boldsymbol{P}} a_u (b = \bigcap_{u \in U}^{\boldsymbol{P}} b_u)$. Then $D_i(a) = \bigcup_{u \in U}^{\boldsymbol{B_P}} D_i(a_u)$, $(D_i(b) = \bigcap_{u \in U}^{\boldsymbol{B_P}} D_i(b_u))$ $i = 1, \cdots, m-1$. By MacNeille's theorem [23] every Boolean algebra, in particular $\boldsymbol{B_P}$, can be extended to a complete Boolean algebra with the preservation of all infinite joins and meets. Let $\boldsymbol{B}$ be such an extension of $\boldsymbol{B_P}$ and let $\boldsymbol{P}'$ be a Post algebra which is a coproduct of $\boldsymbol{B}$ and the chain $\wedge = e_0 \leq \cdots \leq e_{m-1} = \vee$. Then $\boldsymbol{P}'$ is a complete Post algebra of order m which is an extension of $\boldsymbol{P}$ and $\boldsymbol{B_{P'}} = \boldsymbol{B}$. By Epstein's lemma it is easy to see that $a = \bigcup_{u \in U}^{\boldsymbol{P}'} a_u (b = \bigcap_{u \in U}^{\boldsymbol{P}'} b_u)$. Thus any Post algebra $\boldsymbol{P}$ can be extended to a complete Post algebra of the same order with the preservation of all infinite joins and meets in $\boldsymbol{P}$.

It is worth mentioning that the following generalization of the lemma on Boolean algebras (Rasiowa and Sikorski [40], cf. Sec. 3) holds for Post algebras. If in a Post algebra $\boldsymbol{P}$, there is given an enumerable set (S) of infinite joins and meets, then there is a one-to-one homomorphism h of $\boldsymbol{P}$ into a complete Post field $\boldsymbol{P}(X)$ which maps all infinite joins and meets in (S) onto corresponding set-theoretical unions and intersections (an extension of Włodarska [61]).

Let h be a homomorphism of a complete Post algebra $\boldsymbol{P}$ into a complete Post algebra $\boldsymbol{P}'$ of the same order and let $\mathcal{I}$ be an interpretation of all predicates and functors of $\mathcal{L}_m$ in a set $J \neq 0$ and in $\boldsymbol{P}$. Assume that h has the following property: If $\bigcup^{\boldsymbol{P}}_{u \in U} a_u$ ($\bigcap^{\boldsymbol{P}}_{u \in U} b_u$) occurs in the algebraic interpretation $\alpha_{\mathcal{I}}$ of any formula α, then

$$(11) \quad h(\bigcup^{\boldsymbol{P}}_{u \in U} a_u) = \bigcup^{\boldsymbol{P}'}_{u \in U} h(a_u), \quad (h(\bigcap^{\boldsymbol{P}}_{u \in U} b_u) = \bigcap^{\boldsymbol{P}'}_{u \in U} h(b_u)).$$

It is then said that h preserves the infinite joins and meets in $\boldsymbol{P}$ which correspond to quantifiers in the interpretation $\mathcal{I}$. In particular, this is always so if equations (11) hold for all infinite joins and meets in $\boldsymbol{P}$. In such a case the following equations define another interpretation, denoted by $h(\mathcal{I})$, in J and $\boldsymbol{P}'$: $\phi_{h\mathcal{I}} = \phi_{\mathcal{I}}$ for every functor ϕ in $\mathcal{L}_m$; $e_{ih\mathcal{I}} = h(e_{i\mathcal{I}}) = e'_i \in \boldsymbol{P}'$ for $i = 1, \cdots, m-1$; $\rho_{h\mathcal{I}} = h(\rho_{\mathcal{I}})$ for every predicate ρ in $\mathcal{L}_m$. Thus $\mathcal{I}$ and $h\mathcal{I}$ establish the same interpretation of terms. It can easily be proved that, for every formula α,

$$(12) \qquad \alpha_{h\mathcal{I}} = h(\alpha_{\mathcal{I}}).$$

By a model (an algebraic model) of a theory $\mathcal{S}^*_m(\mathcal{A})$ we shall mean any interpretation $\mathcal{I}$ of predicates and functors of this theory in a set $J \neq 0$ and a non-degenerate complete Post algebra $\boldsymbol{P}$ such that, for every α in $\mathcal{A}$, $\alpha_{\mathcal{I}} = \vee$ identically. The following theorem is easy to prove:

For every formula α in a theory $\mathcal{S}^*_m(\mathcal{A})$, if α is a theorem in $\mathcal{S}^*_m(\mathcal{A})$, then $\alpha_{\mathcal{I}} = \vee$ identically in every model $\mathcal{I}$ of $\mathcal{S}^*_m(\mathcal{A})$ in any set $J \neq 0$ and any complete Post algebra $\boldsymbol{P}$.

To prove the converse theorem we introduce the concept of the canonical model of a theory. The canonical interpretation of a consistent theory $\mathcal{S}^*_m(\mathcal{A})$ is the following interpretation $\mathcal{I}_0$ in the set T of all terms and in a complete Post algebra $\boldsymbol{P}$ of order m, which is an extension of the non-degenerate Post algebra $\boldsymbol{A}(\mathcal{S}^*_m(\mathcal{A}))$ preserving all the equations $(\mathrm{Q}_{\mathcal{A}})$: $\phi_{\mathcal{I}_0}(\tau_1, \cdots, \tau_n) = \phi(\tau_1, \cdots, \tau_n)$ for each n-argument functor ϕ in $\mathcal{L}_m$, $n = 0, 1, \cdots$; $\rho_{\mathcal{I}_0}(\tau_1, \cdots, \tau_m) = \| \rho(\tau_1, \cdots, \tau_n) \|_{\mathcal{A}} \in \boldsymbol{P}$, for each n-argument predicate ρ in $\mathcal{L}_m$,

$n = 1, 2, \cdots$. It is proved, as in the case of the intuitionistic predicate calculus $\mathcal{S}_\chi^*$, that, for every formula β in $\mathfrak{L}_m$ and every mapping $s\colon V \to T$,

$$(13) \qquad \beta_{\mathfrak{I}_0}(s) = \| s\beta \|_{\mathfrak{a}} \quad \text{in} \quad \boldsymbol{P},$$

where $s\beta$ denotes the formula obtained from β by the replacement of all occurrences of any free individual variable x in β by the term $s(x)$. In particular, in the case of the identity mapping $s_0\colon V \to T$, we have

$$(14) \qquad \beta_{\mathfrak{I}_0}(s_0) = \| \beta \|_{\mathfrak{a}}.$$

Note that in the proof of (13) the equations $(\mathrm{Q}_{\mathfrak{a}})$ and the fact that they are preserved when $\boldsymbol{A}(\mathcal{S}_m^*(\mathfrak{A}))$ is extended into $\boldsymbol{P}$ plays an essential part. If $\beta \in \mathfrak{A}$, then under the rule of substitution, for every $s\colon V \to T$, $s\beta$ is a theorem of $\mathcal{S}_m^*(\mathfrak{A})$, and hence by (9), $\| s\beta \|_{\mathfrak{a}} = \vee$. This implies with regard to (13) that $\beta_{\mathfrak{I}_0}(s) = \vee$, which proves that $\mathfrak{I}_0$ is a model of $\mathcal{S}_m^*(\mathfrak{A})$ to be called canonical. This model is adequate, for if α is not a theorem in this theory, then by (9) and (14), $\alpha_{\mathfrak{I}_0}(s_0) = \| \alpha \|_{\mathfrak{a}} \neq \vee$. Thus the following theorem holds:

THEOREM 8: *For any formula α in any consistent theory $\mathcal{S}_m^*(\mathfrak{A})$ the following conditions are equivalent:*

(i) $\mathfrak{A} \vdash \alpha$,

(ii) $\alpha_{\mathfrak{I}} = \vee$ *identically in every model $\mathfrak{I}$ of $\mathcal{S}_m^*(\mathfrak{A})$, in any set $J \neq 0$ and any complete non-degenerate Post algebra $\boldsymbol{P}$ of order m,*

(iii) $\alpha_{\mathfrak{I}_0}(s_0) = \vee$ *in a canonical model $\mathfrak{I}_0$ for the identity mapping $s_0\colon V \to T$.*

Models of a theory $\mathcal{S}_m^*(\mathfrak{A})$ in any set $J \neq 0$ and in the m-element Post algebra $\boldsymbol{P}_m$ of order m (which is complete and non-degenerate) are called m-valued. The question arises whether each of the conditions (i), (ii) and (iii) above is equivalent to condition (iv): $\alpha_{\mathfrak{I}} = \vee$ identically in any m-valued model $\mathfrak{I}$ of $\mathcal{S}_m^*(\mathfrak{A})$. The answer is in the affirmative, as it is for $m = 2$, i.e., in the case of theories based on two-valued (classical) logic. The proof is first carried out for those theories $\mathcal{S}_m^*(\mathfrak{A})$ in which the set F_m of formulas is enumerable. In the proof use is made of the following lemma on the existence of D-filters in Post algebras (Rasiowa [37]), which is a generalized version of the Rasiowa-Sikorski lemma (cf. Sec. 3, footnote 14).

LEMMA: *Let* $\boldsymbol{P}$ *be any Post algebra of order m in which the following holds:*

(i) $a_n = \bigcup^{\boldsymbol{P}}_{u \in U_n} a_{nu} \qquad n = 1, 2, \cdots;$
(ii) $b_n = \bigcap^{\boldsymbol{P}}_{v \in V_n} b_{nv} \qquad n = 1, 2, \cdots;$
(iii) $a_0 \neq \vee$.

There is then a prime D-filter ∇ *in* $\boldsymbol{P}$, *which preserves all infinite joins* (i) *and all infinite meets* (ii) *and such that* $a_0 \notin \nabla$. *In other words, in the quotient Post algebra* $\boldsymbol{P}/\nabla$, *which is isomorphic with the m-element Post algebra* $\boldsymbol{P}_m$, *the following holds*: $[a_n] = \bigcup^{\boldsymbol{P}/\nabla}_{u \in U_n} [a_{nu}]$, $n = 1, 2, \cdots$; $[b_n] = \bigcap^{\boldsymbol{P}/\nabla}_{v \in V_n} [b_{nv}]$, $n = 1, 2, \cdots$; $[a_0] \neq \vee$.

It follows from (i), (ii) and Epstein's lemma that

$$D_i(a_n) = \bigcup^{\boldsymbol{B_P}}_{u \in U_n} D_i(a_{nu}), \tag{15}$$

$$D_i(b_n) = \bigcap^{\boldsymbol{B_P}}_{v \in V_n} D_i(b_{nv}), \qquad i = 1, \cdots, m-1.$$

By (iii) and (7) in Sec. 2, $D_{m-1}(a_0) \neq \vee$. Hence $\neg D_{m-1}(a_0) \neq \wedge$ in $\boldsymbol{B_P}$. Since the set of infinite joins and meets which occur in (15) is enumerable and $\neg D_{m-1}(a_0) \neq \wedge$, by the Rasiowa and Sikorski lemma on prime filters in Boolean algebras there is a prime filter ∇_0 in $\boldsymbol{B_P}$ such that $\neg D_{m-1}(a_0) \in \nabla_0$ and so $D_{m-1}(a_0) \notin \nabla_0$, and the said infinite joins and meets are preserved in $\boldsymbol{B_P}/\nabla_0$. It follows therefrom that in $\boldsymbol{B_P}/\nabla_0$, which is the two-element Boolean algebra,

$$[D_i(a_n)] = \bigcup^{\boldsymbol{B_P}/\nabla_0}_{u \in U_n} [D_i(a_{nu})], \tag{16}$$

$$[D_i(b_n)] = \bigcap^{\boldsymbol{B_P}/\nabla_0}_{v \in V_n} [D_i(b_{nv})],$$

for each $i = 1, \cdots, m-1$. By Theorem 4 in Sec. 2, $\nabla = \{a \in \boldsymbol{P}: D_{m-1}(a) \in \nabla_0\}$ is a prime D-filter in $\boldsymbol{P}$ and $\nabla_0 = \boldsymbol{B_P} \cap \nabla$. Since $D_{m-1}(a_0) \notin \nabla_0$, we have $a_0 \notin \nabla$. Making use of (16) and the definition of ∇ it can be proved that the equations (15) are also preserved in $\boldsymbol{B}_{P/\nabla}$, i.e.,

$$[D_i(a_n)] = \bigcup_{u \in U_n}^{\boldsymbol{B}_{P/\nabla}} [D_i(a_{nu})], \quad [D_i(b_n)] = \bigcap_{v \in V_n}^{\boldsymbol{B}_{P/\nabla}} [D_i(b_{nv})],$$

for each $i = 1, \cdots, m - 1$. By reference to Epstein's lemma we can prove easily that the equations (i) and (ii) are preserved in $\boldsymbol{P}/\nabla$.

We now revert to the analysis which preceded the formulation of the above lemma and the outline of its proof. Assume that the set F_m of all formulas in $\mathcal{S}_m^*(\mathfrak{A})$ is enumerable. Let $\boldsymbol{P}$ be a complete Post algebra of order m that is an extension of $\boldsymbol{A}(\mathcal{S}_m^*(\mathfrak{A}))$ such that all infinite joins and meets in $\boldsymbol{A}(\mathcal{S}_m^*(\mathfrak{A}))$ are preserved and let $\mathcal{J}_0$ be the canonical model of $\mathcal{S}_m^*(\mathfrak{A})$ in the set T of all terms and $\boldsymbol{P}$. Assume that α is not a theorem in $\mathcal{S}_m^*(\mathfrak{A})$. Then, by (9), $\| \alpha \|_{\mathfrak{A}} \neq \vee$. Since F_m is enumerable, the set $(Q_{\mathfrak{A}})$ of infinite joins and meets in $\boldsymbol{A}(\mathcal{S}_m^*(\mathfrak{A}))$ that correspond to quantifiers is enumerable, too. By the lemma above there is a prime D-filter ∇ in $\boldsymbol{P}$ which preserves the infinite joins and meets $(Q_{\mathfrak{A}})$ in $\boldsymbol{P}$ and such that $\| \alpha \|_{\mathfrak{A}} \notin \nabla$. Let h be the natural epimorphism from $\boldsymbol{P}$ onto $\boldsymbol{P}/\nabla$, i.e., for every $a \in \boldsymbol{P}$, $h(a) = [a]$ in $\boldsymbol{P}/\nabla$. In particular, for every formula β in $\mathcal{S}_m^*(\mathfrak{A})$, $h(\| \beta \|_{\mathfrak{A}}) = [\| \beta \|_{\mathfrak{A}}]$. Then h preserves all infinite joins and meets in $\boldsymbol{P}$ which correspond to quantifiers in the interpretation $\mathcal{J}_0$ (the canonical model $\mathcal{J}_0$ of $\mathcal{S}_m^*(\mathfrak{A})$). Hence, by (12), for the interpretation $h\mathcal{J}_0$ in T and $\boldsymbol{P}/\nabla$, and any formula β we have $\beta_{h\mathcal{J}_0} = h\beta_{\mathcal{J}_0}$. Since $\mathcal{J}_0$ is a model of $\mathcal{S}_m^*(\mathfrak{A})$, for every β in $\mathfrak{A}$, $\beta_{\mathcal{J}_0} = \vee$ identically. Thus $\beta_{h\mathcal{J}_0} = h\beta_{\mathcal{J}_0} = h(\vee) = \vee$. This proves that $h\mathcal{J}_0$ is an m-valued model of $\mathcal{S}_m^*(\mathfrak{A})$. On the other hand, on account of (14) and the fact that $\| \alpha \|_{\mathfrak{A}} \notin \nabla$, we obtain $\alpha_{h\mathcal{J}_0}(s_0) = h\alpha_{\mathcal{J}_0}(s_0) = h \| \alpha \|_{\mathfrak{A}} = [\| \alpha \|_{\mathfrak{A}}] \neq \vee$. We have thus proved the following

THEOREM 9: *If the set F_m of formulas of a theory $\mathcal{S}_m^*(\mathfrak{A})$ is enumerable, and if α is not a theorem in $\mathcal{S}_m^*(\mathfrak{A})$, then there is an m-valued model $\mathcal{J}$ of that theory in the enumerable set T of all terms such that $\alpha_{\mathcal{J}}$ is not identically equal to $\vee$.*

In Theorem 9, the condition stating that F_m is enumerable can be replaced by one stating that $\mathfrak{A}$ is at most enumerable. For assume that $\mathfrak{A}$ is at most enumerable and that α is not a theorem in $\mathcal{S}_m^*(\mathcal{A})$. Consider a theory $\mathcal{S}_m^{*\prime}(\mathcal{A})$ in which the set F_m' of formulas is enumerable and included in F_m and such that it contains $\mathcal{A}$ and α. Obviously, α *a fortiori* is not a theorem in $\mathcal{S}_m^{*\prime}(\mathcal{A})$. By applying Theorem 9, we infer that there is an m-valued model $\mathcal{J}'$ of $\mathcal{S}_m^{*\prime}(\mathcal{A})$ (in the enumerable set T' of terms in $\mathcal{S}_m^{*\prime}(\mathcal{A})$) such that $\alpha_{\mathcal{J}'}$ is not identically equal to $\vee$. If we extend $\mathcal{J}'$ arbitrarily by establishing an interpretation in T' and $\boldsymbol{P}_m$ of those predicates and functors of $\mathcal{S}_m^*(\mathcal{A})$ which do not occur in $\mathcal{S}_m^{*\prime}(\mathcal{A})$, then we obtain an m-valued model $\mathcal{J}$ of $\mathcal{S}_m^*(\mathcal{A})$ in T', and then it turns out that $\alpha_{\mathcal{J}}$ is not identically equal to $\vee$. This is due to the fact that the interpretations $\mathcal{J}'$ and $\mathcal{J}$ coincide in the set of the formulas of $\mathcal{S}_m^{*\prime}(\mathcal{A})$ (which contains $\mathcal{A}$ and α). Hence we obtain the following theorem which is a generalized version of the Löwenheim-Skolem-Gödel theorem for the classical predicate calculi.

THEOREM 10: *If the set $\mathcal{A}$ of specific axioms of a consistent theory $\mathcal{S}_m^*(\mathcal{A})$ is at most enumerable, then for every formula α of that theory the following conditions are equivalent:*

(i) $\mathcal{A} \vdash \alpha$,
(ii) $\alpha_{\mathcal{J}} = \vee$ *identically in every model $\mathcal{J}$ of that theory, in any set $J \neq 0$ and any complete Post algebra $\boldsymbol{P}$ of order m,*
(iii) $\alpha_{\mathcal{J}} = X$ *identically in every model $\mathcal{J}$ of $\mathcal{S}_m^*(\mathcal{A})$, in any set $J \neq 0$ and any complete Post field $\boldsymbol{P}(X)$ of order m,*
(iv) $\alpha_{\mathcal{J}_0}(s_0) = \vee$ *in a canonical model $\mathcal{J}_0$ of $\mathcal{S}_m^*(\mathcal{A})$ for the identity mapping $s_0\colon V \to T$,*
(v) $\alpha_{\mathcal{J}} = \vee$ *identically in every m-valued model $\mathcal{J}$ of $\mathcal{S}_m^*(\mathcal{A})$, in an enumerable set J.*

Indeed, we just proved that (v) implies (i). By Theorem 8 the conditions (i), (ii), (iv) are equivalent. Clearly (ii) implies (iii), and (iii) implies (v), because $\boldsymbol{P}_m$ is isomorphic with a complete Post field $\boldsymbol{P}(X)$ of order m.

In particular, Theorem 10 is applicable if $\mathcal{A} = 0$, i.e., to an m-valued predicate calculus $\mathcal{S}_m^*$. In fact, it follows from the definition of an algebraic model of a theory $\mathcal{S}_m^*(0)$ that every interpretation $\mathcal{J}$ in any set $J \neq 0$ and in any non-degenerate complete Post algebra $\boldsymbol{P}$ of order m is a model of that theory. On the other hand, if $\mathcal{S}_m^*(\mathcal{A})$ has a model, then it is consistent. This is so because if, for a formula α, α is a theorem of $\mathcal{S}_m^*(\mathcal{A})$, then in every model $\mathcal{J}$ of it, $\alpha_{\mathcal{J}} = \vee$ identically and $\neg\alpha_{\mathcal{J}} = \wedge$. Hence $\neg\alpha$ is not a theorem in that theory, and thus the theory is consistent. This, taken together with the previous comment, implies that $\mathcal{S}_m^* = \mathcal{S}_m^*(0)$ is consistent. Theorem 10, when applied to $\mathcal{S}_m^*$, takes on the following form which is a generalized version of the Gödel completeness theorem:

THEOREM 11: *For every formula α in an m-valued predicate calculus $\mathcal{S}_m^*$ the following conditions are equivalent:*

(i) $\vdash \alpha$,

(ii) $\alpha_{\mathcal{J}} = \vee$ *identically for every interpretation $\mathcal{J}$ in any set $J \neq 0$ and in any complete Post algebra **P** of order m,*

(iii) $\alpha_{\mathcal{J}} = X$ *identically for every interpretation $\mathcal{J}$ in any set $J \neq 0$ and in any complete Post field $\boldsymbol{P}(X)$ of order m,*

(iv) $\alpha_{\mathcal{J}_0}(s_0) = \vee$ *for a canonical model $\mathcal{J}_0$ of $\mathcal{S}_m^*(0)$ and the identity mapping s_0: $V \to T$,*

(v) $\alpha_{\mathcal{J}} = \vee$ *identically for every interpretation $\mathcal{J}$ in an enumerable set J and in the m-element Post algebra $\boldsymbol{P}_m$ or order m.*

One of the fundamental concepts in the theory of models of classical elementary formalized theories is that of the ultraproduct of semantic models (Łoś [19]). It can be generalized so as to cover m-valued models (Rasiowa [38]).

Let N be a non-empty set and let, for every $n \in N$, $\mathcal{J}_n$ be an interpretation of the functors and predicates in $\mathcal{L}_m$ in a set $J_n \neq 0$ and in a complete Post algebra $\boldsymbol{P}^{(n)}$ of order m. Since the class of all Post algebras of order m is equationally definable, hence the product $\boldsymbol{P} = \prod_{n \in N} \boldsymbol{P}^{(n)}$ is a Post algebra of order m. On the other hand, it is known that a product of complete distributive lattices is a complete distributive lattice. Hence $\boldsymbol{P}$ is a complete Post algebra of order m. The product of interpretations $\prod_{n \in N} \mathcal{J}_n$ is an interpretation $\mathcal{J}$ in the set $\boldsymbol{J} = \prod_{n \in N} J_n$ and in the Post algebra $\boldsymbol{P} = \prod_{n \in N} \boldsymbol{P}^{(n)}$ defined thus: for every $j_1 = (j_{1n})_{n \in N}, \cdots, j_k = (j_{kn})_{n \in N}$, which are elements in $\boldsymbol{J}$,

$\phi_{\mathcal{J}}(j_1, \cdots, j_k) = (\phi_{\mathcal{J}_n}(j_{1n}, \cdots, j_{kn}))_{n \in N}$ for every k-argument functor ϕ in $\mathfrak{L}_m$, $k = 0, 1, \cdots,$

$\rho_{\mathcal{J}}(j_1, \cdots, j_k) = (\rho_{\mathcal{J}_n}(j_{1n}, \cdots, j_{kn}))_{n \in N}$ for every k-argument predicate ρ in $\mathfrak{L}_m$, $k = 1, 2, \cdots$.

Note that every mapping $v: V \to J$ determines, for every $n \in N$, a mapping $v^n: V \to J_n$ such that $v(x) = (v^n(x))_{n \in N}$ for every free individual variable $x \in V$. Conversely, if, for every n, v^n maps V into J_n, then $v(x) = (v^n(x))_{n \in N}$ defines a mapping $v: V \to J$. It can easily be proved that, for every formula α and for every $v: V \to J$,

(17) $$\alpha_{\mathcal{J}}(v) = (\alpha_{\mathcal{J}_n}(v^n))_{n \in N}.$$

Consider now the case when, for every $n \in N$, $\mathcal{J}_n$ is an interpretation in $J_n \neq 0$ and in the m-element Post algebra $\boldsymbol{P}_m$ of order m. Consider for that case any prime filter ∇ in the field $\boldsymbol{B}(N)$ of all subsets of N. Let $\boldsymbol{P} = \prod_{n \in N} \boldsymbol{P}^{(n)}$, where $\boldsymbol{P}^{(n)} = \boldsymbol{P}_m$ for each $n \in N$, and let $\boldsymbol{B}_{\boldsymbol{P}}$ be the Boolean algebra that corresponds to the Post algebra $\boldsymbol{P}$. In this case $\boldsymbol{B}_{\boldsymbol{P}} = \prod_{n \in N} \boldsymbol{B}_{\boldsymbol{P}^{(n)}}$, where, for each $n \in N$, $\boldsymbol{B}_{\boldsymbol{P}^{(n)}} = \boldsymbol{B}_{\boldsymbol{P}_m} = \boldsymbol{B}_0$ is the two-element Boolean algebra $\boldsymbol{B}_0$. $\boldsymbol{B}_{\boldsymbol{P}}$ being the product $\prod_{n \in N} \boldsymbol{B}_{\boldsymbol{P}^{(n)}}$ of two-element Boolean algebras, is isomorphic with the field $\boldsymbol{B}(N)$ of all subsets of N, the required isomorphism h_0 being defined as follows: $h_0(a_n)_{n \in N} = \{n \in N: a_n = \vee\}$ for every $(a_n)_{n \in N} \in \boldsymbol{B}_{\boldsymbol{P}}$. The mapping h_1 of the field $\boldsymbol{B}(N)$ onto the two-element Boolean algebra $\boldsymbol{B}(N)/\nabla$, which is defined thus: $h_1(Y) = \vee$ if $Y \in \nabla$ and $h_1(Y) = \wedge$ if $Y \notin \nabla$ (for every $Y \in \boldsymbol{B}(N)$), is an epimorphism. Let $g(a_n)_{n \in N}$ stand for $h_1(h_0(a_n)_{n \in N})$. It follows from this definition of the mapping g that

(18) $$g(a_n)_{n \in N} = \vee \quad \text{if} \quad \{n \in N: a_n = \vee\} \in \nabla,$$
$$g(a_n)_{n \in N} = \wedge \quad \text{if} \quad \{n \in N: a_n = \vee\} \notin \nabla.$$

It can be proved that g is an epimorphism of $\boldsymbol{B}_{\boldsymbol{P}}$ onto the two-element Boolean algebra such that it preserves all infinite joins and meets in $\boldsymbol{B}_{\boldsymbol{P}}$ (cf. Rasiowa and Sikorski [41]). The mapping g

can be extended to an epimorphism h of the Post algebra $\boldsymbol{P}$ onto the m-element Post algebra $\boldsymbol{P}_m$ thus: for every element $a = (D_1(a) \cap e_1) \cup \cdots \cup (D_{m-1}(a) \cap e_{m-1})$ in $\boldsymbol{P}$, $h(a) = (g(D_1(a)) \cap e_1) \cup \cdots \cup (g(D_{m-1}(a)) \cap e_{m-1})$. It can be proved that h preserves all infinite joins and meets in $\boldsymbol{P}$; to do so one has to refer to Epstein's lemma and the fact that g preserves all infinite joins and meets in $\boldsymbol{B_P}$.

The ultraproduct $\mathcal{I} = (\prod_{n \in N} \mathcal{I}_n)/\nabla$ of the interpretations $\mathcal{I}_n$, $n \in N$, determined by ∇ is now defined thus: let $\mathcal{I}_0 = \prod_{n \in N} \mathcal{I}_n$, then

$$(19) \qquad\qquad \mathcal{I} = h\mathcal{I}_0.$$

The ultraproduct $\mathcal{I}$ is by definition an interpretation in the set $J = \prod_{n \in N} J_n$ and in the m-element Post algebra $\boldsymbol{P}_m$ of order m. Another, equivalent definition of $\mathcal{I}$ is as follows: for any

$j_1 = (j_{1n})_{n \in N}, \cdots, j_k = (j_{kn})_{n \in N}$ in J,

$\phi_{\mathcal{I}}(j_1, \cdots, j_k) = (\phi_{\mathcal{I}_n}(j_{1n}, \cdots, j_{kn}))_{n \in N}$ for each k-argument functor ϕ in $\mathcal{L}_m$, $k = 0, 1, \cdots$,

$D_i(\rho_{\mathcal{I}}(j_1, \cdots, j_k))$

$$= \begin{cases} \vee & \text{if} \quad \{n \in N : D_i(\rho_{\mathcal{I}_n}(j_{1n}, \cdots, j_{kn})) = \vee\} \in \nabla \\ \wedge & \text{otherwise} \end{cases}$$

for each k-argument predicate ρ in $\mathcal{L}_m$, $k = 1, 2, \cdots$; $i = 1, \cdots, m - 1$,

$\rho_{\mathcal{I}}(j_1, \cdots, j_k) = (D_1(\rho_{\mathcal{I}}(j_1, \cdots, j_k)) \cap e_1) \cup \cdots \cup (D_{m-1}(\rho_{\mathcal{I}}(j_1, \cdots, j_k)) \cap e_{m-1})$ for every k-argument predicate ρ in $\mathcal{L}_m$, $k = 1, 2, \cdots$, where $e_1, \cdots, e_{m-1}$ belong to $\boldsymbol{P}_m$.

The following is a generalized version of Łoś's theorem [19] for ultraproducts of semantic interpretations of classical predicate calculi.

THEOREM 12: *If $\mathcal{I}$ is the ultraproduct of interpretations $\mathcal{I}_n$ in J_n, $n \in N$, determined by ∇, then for every formula α of $\mathcal{L}_m$, every* *mapping* $v: V \to J$, *where* $J = \prod_{n \in N} J_n$, *and every* $i = 1, \cdots, m-1$,

$$D_i(\alpha_{\mathcal{J}}(v)) = \begin{cases} \vee & \text{if } \{n \in N: D_i(\alpha_{\mathcal{J}_n}(v^n)) = \vee\} \in \nabla \\ \wedge & \text{otherwise.} \end{cases}$$

In fact, by (19), (12), the definition of $\mathcal{J}_0$, (17) and the definition of h we have: $D_i(\alpha_{\mathcal{J}}(v)) = D_i(\alpha_{h\mathcal{J}_0}(v)) = D_i(h(\alpha_{\mathcal{J}_0}(v))) = D_i(h(\alpha_{\mathcal{J}_n}(v^n))_{n \in N} = h(D_i(\alpha_{\mathcal{J}_n}(v^n))_{n \in N}) = g(D_i(\alpha_{\mathcal{J}_n}(v^n))_{n \in N})$. This and (18) yield the required relationships.

Scott's idea (cf. Rasiowa and Sikorski [41]) of the proof of the Löwenheim-Skolem-Gödel-Malcev theorem on the existence of semantic models for any elementary consistent theories based on classical logic will be used to prove a generalized version of this theorem for theories based on m-valued logics. In the proof reference will be made to Theorem 12.

Assume that a theory $\mathcal{S}_m^*(\mathfrak{A})$ is consistent, and let N stand for the family of all finite subsets of $\mathfrak{A}$. For every formula $\alpha \in \mathfrak{A}$, let N_α denote the set of $n \in N$ such that $\alpha \in n$. For any $\alpha_1, \cdots, \alpha_k \in \mathfrak{A}$, the intersection $N_{\alpha_1} \cap \cdots \cap N_{\alpha_k} \neq 0$, because $n = \{\alpha_1, \cdots, \alpha_k\}$ belongs to this set. This guarantees that the filter ∇_0 in the field $\boldsymbol{B}(N)$ of all subsets of N, generated by all N_α, $\alpha \in \mathfrak{A}$, is a proper one. By (5) in Sec. 1, there is then a prime filter ∇ in $\boldsymbol{B}(N)$ which contains ∇_0. On account of Theorem 10, for every theory $\mathcal{S}_m^*(n)$, $n \in N$, there is an m-valued model $\mathcal{J}_n$ in a set $J_n \neq 0$. Let $\mathcal{J} = (\prod_{n \in N} \mathcal{J}_n)/\nabla$. The interpretation $\mathcal{J}$ is an m-valued model of the theory $\mathcal{S}_m^*(\mathfrak{A})$. For assume that $\alpha \in \mathfrak{A}$. In this case, for every $v: V \to J$, where $J = \prod_{n \in N} J_n$, we have $N_\alpha = \{n \in N\colon \alpha \in n\} \subset \{n \in N\colon \alpha_{\mathcal{J}_n}(v^n) = \vee\}$, as the conditions: $\alpha \in n$, and $\mathcal{J}_n$ is an m-valued model of $\mathcal{S}_m^*(n)$, imply $\alpha_{\mathcal{J}_n}(v^n) = \vee$. On the other hand, $N_\alpha \in \nabla_0 \subset \nabla$. Hence every subset of N that includes N_α, in particular $\{n \in N\colon \alpha_{\mathcal{J}_n}(v^n) = \vee\}$, is in ∇. But in the Post algebra $\boldsymbol{P}_m$, for every element a, the condition $a = \vee$ is equivalent to the condition $D_{m-1}(a) = \vee$. Hence $\alpha_{\mathcal{J}_n}(v^n) = \vee$ iff $D_{m-1}(\alpha_{\mathcal{J}_n}(v^n)) = \vee$. Consequently, $\{n \in N: D_{m-1}(\alpha_{\mathcal{J}_n}(v^n)) = \vee\}$ is in ∇, too. This and Theorem 12 yield $D_{m-1}(\alpha_{\mathcal{J}}(v)) = \vee$ in $\boldsymbol{P}_m$, which implies $\alpha_{\mathcal{J}}(v) = \vee$. Since this condition is satisfied for every $\alpha \in \mathfrak{A}$ and every mapping $v\colon V \to J$, it is to be concluded that $\mathcal{J}$ is an m-valued model of $\mathcal{S}_m^*(\mathfrak{A})$. We have thus proved the following

THEOREM 13: *Every consistent theory* $\mathcal{S}_m^*(\mathfrak{A})$ *has an* m*-valued model.*

The above theorem easily yields the theorem stating that if a formula α is not a theorem of $\mathcal{S}_m^*(\mathfrak{A})$, then there is an m-valued model $\mathcal{J}$ of that theory such that $\alpha_{\mathcal{J}}$ does not identically equal $\vee$, which makes it possible to add in Theorem 8 two conditions, (iv) and (v), each of which is equivalent to each of the preceding ones:

(iv) $\alpha_{\mathcal{J}} = X$ identically for every model $\mathcal{J}$ of $\mathcal{S}_m^*(\mathfrak{A})$ in every complete (non-degenerate) Post field $\boldsymbol{P}(X)$ of order m and every set $J \neq 0$,

(v) $\alpha_{\mathcal{J}} = \vee$ identically for every m-valued model $\mathcal{J}$ of $\mathcal{S}_m^*(\mathfrak{A})$ in every set $J \neq 0$.

The proof that condition (iv) is equivalent with each other is carried out similarly as in Theorem 10.

The following theorem on ultraproducts of m-valued models is worth mentioning:

THEOREM 14: *If, for every* $n \in N$, *where* N *is an abstract set,* $\mathcal{J}_n$ *is an* m*-valued model of* $\mathcal{S}_m^*(\mathfrak{A})$ *in* $J_n \neq 0$ *and* ∇ *is a prime filter in the field* $\boldsymbol{B}(N)$ *of all subsets of* N, *then the ultraproduct* $\mathcal{J} = (\prod_{n\in N} \mathcal{J}_n)/\nabla$ *is an* m*-valued model of* $\mathcal{S}_m^*(\mathfrak{A})$.

For assume that $\alpha \in \mathfrak{A}$. Then, by the rule of inference (r_m), $\mathfrak{A} \vdash D_{m-1}(\alpha)$. Let $v\colon V \to J$, where $J = \prod_{n\in N} J_n$, and $v(x) = (v^n(x))_{n\in N}$ for every free individual variable $x \in V$. Since $\mathcal{J}_n$ is a model of $\mathcal{S}_m^*(\mathfrak{A})$, $n \in N$, and $\mathfrak{A} \vdash D_{m-1}(\alpha)$, we infer that $D_{m-1}(\alpha_{\mathcal{J}_n}(v^n)) = \vee$ for each $n \in N$. Thus $\{n \in N\colon D_{m-1}(\alpha_{\mathcal{J}_n}(v^n)) = \vee\} = N \in \nabla$, because in every Boolean algebra, in particular in $\boldsymbol{B}(N)$, the unit element belongs to every filter. Hence, by Theorem 12, $D_{m-1}(\alpha_{\mathcal{J}}(v)) = \vee$ in $\boldsymbol{P}_m$, which implies that $\alpha_{\mathcal{J}}(v) = \vee$.

Other well-known theorems on elementary classical formalized theories can also be generalized so as to cover elementary formalized theories based on m-valued logics. A few examples of such theorems will be given below.

THE DEDUCTION THEOREM: *Let α be a formula without free individual variable in $\mathcal{S}_m^*(\mathfrak{A})$. Then a formula β is a theorem in $\mathcal{S}_m^*(\mathfrak{A} \cup \{\alpha\})$ iff $(D_{m-1}(\alpha) \Rightarrow \beta)$ is a theorem in $\mathcal{S}_m^*(\mathfrak{A})$.*

Assume that $\mathfrak{A} \vdash (D_{m-1}(\alpha) \Rightarrow \beta)$. This implies $\mathfrak{A} \cup \{a\} \vdash (D_{m-1}(\alpha) \Rightarrow \beta)$. Since α is an axiom of $\mathcal{S}_m^*(\mathfrak{A} \cup \{\alpha\})$, by rule (r_m), $D_{m-1}(\alpha)$ is a theorem of this theory. Thus $\mathfrak{A} \cup \{\alpha\} \vdash D_{m-1}(\alpha)$. On applying modus ponens we have $\mathfrak{A} \cup \{\alpha\} \vdash \beta$, which proves the sufficiency of the condition under consideration. If $(D_{m-1}(\alpha) \Rightarrow \beta)$ is not a theorem in $\mathcal{S}_m^*(\mathfrak{A})$, then there is an m-valued model $\mathfrak{J}$ of that theory in a set $J \neq 0$ and a mapping $v: V \to J$ such that $D_{m-1}(\alpha_{\mathfrak{J}}(v)) \Rightarrow \beta_{\mathfrak{J}}(v) = (D_{m-1}(\alpha) \Rightarrow \beta)_{\mathfrak{J}}(v) \neq \vee$. Since free individual variables do not occur in α, $D_{m-1}(\alpha_{\mathfrak{J}}(v))$ has a constant value in $\boldsymbol{P}_m$, which is independent of v. In $\boldsymbol{P}_m$ for every element a we have $D_{m-1}(a) = \vee$ or $D_{m-1}(a) = \wedge$. The case $D_{m-1}(\alpha_{\mathfrak{J}}(v)) = \wedge$ is impossible, because then we would have $D_{m-1}(\alpha_{\mathfrak{J}}(v)) \Rightarrow \beta_{\mathfrak{J}}(v) = \vee$. It follows accordingly from the above that $D_{m-1}(\alpha_{\mathfrak{J}}(v)) = \vee$, so that $\alpha_{\mathfrak{J}}(v) = \vee$, which proves that $\mathfrak{J}$ is an m-valued model of $\mathcal{S}_m^*(\mathfrak{A} \cup \{\alpha\})$. In this model $\beta_{\mathfrak{J}}(v) \neq \vee$, since otherwise we would have $D_{m-1}(\alpha_{\mathfrak{J}}(v)) \Rightarrow \beta_{\mathfrak{J}}(v) = \vee$. By Theorem 8 we infer that β is not a theorem in $\mathcal{S}_m^*(\mathfrak{A} \cup \{\alpha\})$, which concludes the proof of the deduction theorem.

Let now F_m^0 be the set of open formulas in F_m, i.e., those formulas in which no quantifiers occur. A theory $\mathcal{S}_m^*(\mathfrak{A})$ is called open if $\mathfrak{A} \subset F_m^0$. The theorem on the elimination of quantifiers in formal proofs holds for the open theories based on m-valued logics as it does for the classical open theories (the first ϵ-theorem, Hilbert and Bernays [14]). The theorem states that an open formula is a theorem in an open theory $\mathcal{S}_m^*(\mathfrak{A})$ iff it can be proved in that theory in such a way that no quantifiers occur in the proof. Here is a more precise formulation of this theorem. Let $\mathfrak{A}_{ml}^0$ be the set of all those logical axioms of $\mathcal{S}_m^*$ which are open formulas, and let $\mathfrak{A}$ be a set of open formulas. Then $\mathcal{S}_m^{*0}(\mathfrak{A})$ will stand for a theory in which $\mathfrak{A}_{ml}^0$ is the set of logical axioms, and in which only three rules of inference are adopted: the *modus ponens*, (r_m), and the rule of substitution for free individual variables. Under these assumptions, the following theorem holds:

THEOREM 15: *An open formula* α *is a theorem in an open theory* $\mathcal{S}_m^*(\mathfrak{A})$ *iff it is a theorem in* $\mathcal{S}_m^{*0}(\mathfrak{A})$ [16].

The sufficiency of this condition is self-evident. To prove the necessity assume that α is not a theorem in $\mathcal{S}_m^{*0}(\mathfrak{A})$. We then construct an algebra $\boldsymbol{A}(\mathcal{S}_m^{*0}(\mathfrak{A}))$ of the theory $\mathcal{S}_m^{*0}(\mathfrak{A})$ in a manner analogous to that in which an algebra $\boldsymbol{A}(\mathcal{S}_m^{*}(\mathfrak{A}))$ is constructed, and prove that it is a non-degenerate Post algebra of order m, and that, for any formula $\beta \in F_m^0$, $\| \beta \|_{\mathfrak{A}} = \vee$ iff β is a theorem in $\mathcal{S}_m^{*0}(\mathfrak{A})$. Hence $\| \alpha \|_{\mathfrak{A}} \neq \vee$. The algebra $\boldsymbol{A}(\mathcal{S}_m^{*0}(\mathfrak{A}))$ is then extended into a complete Post algebra $\boldsymbol{P}$ of order m, and next we consider a canonical interpretation $\mathcal{J}_0$ of $\mathcal{S}_m^{*0}(\mathfrak{A})$ in the set T of all terms and in $\boldsymbol{P}$. No restriction concerning the preservation of infinite joins and meets by extending $\boldsymbol{A}(\mathcal{S}_m^{*0}(\mathfrak{A}))$ to $\boldsymbol{P}$ is made, because no equation in $(Q_{\mathfrak{A}})$ holds in $\boldsymbol{A}(\mathcal{S}_m^{*0}(\mathfrak{A}))$. For every open formula β the condition $\beta_{\mathcal{J}_0}(s) = \| s\beta \|_{\mathfrak{A}}$ is then satisfied, where $s\colon V \to T$, and $s\beta$ is a formula obtained from β by the replacement of every free variable x by the term $s(x)$. In particular, if $\beta \in \mathfrak{A}$, then $s\beta$ is a theorem in $\mathcal{S}_m^{*0}(\mathfrak{A})$, so that $\| s\beta \|_{\mathfrak{A}} = \vee$, and consequently $\beta_{\mathcal{J}_0}(s) = \vee$, which proves that $\mathcal{J}_0$ is a model of $\mathcal{S}_m^{*}(\mathfrak{A})$. On the other hand, for the identity mapping $s_0\colon V \to T$ we have $\alpha_{\mathcal{J}_0}(s_0) = \| \alpha \|_{\mathfrak{A}} \neq \vee$. By Theorem 8, α is not a theorem in $\mathcal{S}_m^{*}(\mathfrak{A})$, which completes the proof of Theorem 15.

A formula α in $\mathfrak{L}_m$ is said to be prenex provided that all the quantifiers which occur in it are at the beginning of the formula and are followed by an expression without quantifiers. A prenex formula α is said to be a prenex form of a formula β provided that both implications $(\alpha \Rightarrow \beta)$ and $(\beta \Rightarrow \alpha)$ are theorems in $\mathcal{S}_m^*$. It is proved that, as in the case of the classical predicate calculi, every formula β in $\mathfrak{L}_m$ has a prenex form α.

Open theories based on m-valued logics have an interesting

[16] The proof of Theorem 15 is quite analogous to that for the classical open theories (Rasiowa [35]).

property, formulated in a generalized version of the Herbrand theorem for the classical open theories. Roughly speaking, it states that, for every prenex closed formula α (i.e., without free individual variables) in an open theory $\mathcal{S}_m^*(\mathfrak{A})$, it is possible to construct a sequence of open formulas $H_1, H_2, \cdots$ (Herbrand's disjunctions) so that α is a theorem in $\mathcal{S}_m^*(\mathfrak{A})$ iff at least one of the formulas $H_1, H_2, \cdots$ is a theorem. This theorem will be formulated in a precise version for a special case of α, the generalized version being left for the reader.

THEOREM 16: *Let $\mathcal{S}_m^*(\mathfrak{A})$ be an open theory in which $V = \{x_1, x_2, \cdots\}$ and the set F_m of all formulas is enumerable. Let $\tau_1, \tau_2, \cdots$ be a sequence of terms in that theory such that every term occurs exactly once in that sequence. Let α be a prenex closed formula in the form*

$$(20) \qquad \bigcup\xi_1 \bigcap\eta_1 \bigcup\xi_2 \bigcap\eta_2 \bigcup\xi_3\beta(\xi_1, \eta_1, \xi_2, \eta_2, \xi_3).$$

Let k be a one-to-one mapping of the set N of positive integers into itself such that the free variable $x_{k(i)}$ does not occur in the term τ_i for each $i \in N$. Let further p: $N \times N \to N$ be a one-to-one mapping that satisfies the condition $p(N \times N) \cap k(N) = 0$ and the condition stating that the free variable $x_{p(i,j)}$ occurs neither in τ_i nor in τ_j for $i, j \in N$. We put for each $n \in N$

$$(21) \qquad H_n = \bigcup_{i=1}^{n} \bigcup_{j=1}^{n} \bigcup_{h=1}^{n} \beta(\tau_i, x_{k(i)}, \tau_j, x_{p(i,j)}, \tau_h)$$

where $\bigcup_{i=1}^{n} \gamma_i$ stands for $(\cdots(\gamma_1 \cup \gamma_2) \cup \cdots \cup \gamma_n)$. In such a case, α is a theorem in $\mathcal{S}_m^*(\mathfrak{A})$ iff there is an $n \in N$ such that H_n is a theorem in that theory.

The proof of this theorem (due to E. Perkowska[17]) resembles that for the classical open theories [18]. In particular, the proof of the sufficiency of the condition under consideration is based on the technique of proving α if H_n is proved for a certain n, while the proof of the necessity of this condition is algebraic in nature and

---

[17] "Herbrand Theorem . . .", Bull. Acad. Pol. Sci., Ser. Sci. Math. Astron. Phys., vol. 19 (1971), pp. 893–899.

[18] Due to Łoś, Rasiowa and Mostowski (1956), cf. [41].

is based on the following idea. Assume that for each $n \in N$, H_n is not a theorem in $\mathcal{S}_m^*(\mathfrak{A})$. Then H_n is not a theorem in $\mathcal{S}_m^{*0}(\mathfrak{A})$, for $n \in N$ (see Theorem 15). We then construct the algebra $\boldsymbol{P} = \boldsymbol{A}(\mathcal{S}_m^{*0}(\mathfrak{A}))$ of the theory $\mathcal{S}_m^{*0}(\mathfrak{A})$, which is a non-degenerate Post algebra of order m. In that algebra $\| H_n \|_{\mathfrak{A}} \neq \vee$, for $n \in N$, which implies by (7) in Sec. 2 that $D_{m-1} \| H_n \|_{\mathfrak{A}} \neq \vee$, $n \in N$. Consequently, $\neg D_{m-1} \| H_n \|_{\mathfrak{A}} \neq \wedge$, $n \in N$, in the Boolean algebra $\boldsymbol{B_P}$. Let ∇_0 be the filter in $\boldsymbol{B_P}$ generated by all $\neg D_{m-1} \| H_n \|_{\mathfrak{A}}$, $n \in N$. Thus $\neg D_{m-1} \| H_n \|_{\mathfrak{A}} \in \nabla_0$ for each $n \in N$. It can easily be proved that ∇_0 is a proper filter in $\boldsymbol{B_P}$ and therefore is contained in a prime filter ∇^* (see (5) in Sec. 1). Thus $\neg D_{m-1} \| H_n \|_{\mathfrak{A}} \in \nabla^*$, $n \in N$, which implies that $D_{m-1} \| H_n \|_{\mathfrak{A}} \notin \nabla^*$ for each $n \in N$. Then $\nabla = \{\| \gamma \|_{\mathfrak{A}} \in \boldsymbol{P}\colon D_{m-1}(\| \gamma \|_{\mathfrak{A}}) \in \nabla^*\}$ is, by Theorem 4 in Sec. 2, a prime D-filter in $\boldsymbol{P}$ and $\| H_n \|_{\mathfrak{A}} \notin \nabla$ for each $n \in N$. This implies that in the quotient algebra $\boldsymbol{P}/\nabla$, which is isomorphic with the m-element Post algebra $\boldsymbol{P_m}$, $[\| H_n \|_{\mathfrak{A}}] \neq \vee$. This and the form of H_n yield that

$$(22) \quad [\| \beta(\tau_i, x_{k(i)}, \tau_j, x_{p(i,j)}, \tau_h) \|_{\mathfrak{A}}] \neq \vee \quad \text{for all} \quad i, j, h \in N.$$

On the other hand, an interpretation $\mathcal{I}_0$ in the set T of all terms, and in $\boldsymbol{P}/\nabla$ defined thus: $\phi_{\mathcal{I}_0}(\sigma_1, \cdots, \sigma_n) = \phi(\sigma_1, \cdots, \sigma_n)$, for each n-argument functor ϕ, $n = 0, 1, \cdots$, and any terms $\sigma_1, \cdots, \sigma_n$ in T, and $\rho_{\mathcal{I}_0}(\sigma_1, \cdots, \sigma_n) = [\| \rho(\sigma_1, \cdots, \sigma_n) \|_{\mathfrak{A}}]$ for each n-argument predicate ρ, $n = 1, 2, \cdots$, and any terms $\sigma_1, \cdots, \sigma_n$ in T, is an m-valued model of $\mathcal{S}_m^*(\mathfrak{A})$. This is so because, for every open formula γ and for every $s\colon V \to T$, the condition $\gamma_{\mathcal{I}_0}(s) = [\| s\gamma \|_{\mathfrak{A}}]$ is satisfied. In particular, if $\gamma \in \mathfrak{A}$, then γ is an open formula and $s\gamma$ is a theorem in $\mathcal{S}_m^{*0}(\mathfrak{A})$, which implies that $\| s\gamma \|_{\mathfrak{A}} = \vee$, and hence that $[\| s\gamma \|_{\mathfrak{A}}] = \vee$. Consequently, $\gamma_{\mathcal{I}}(s) = \vee$. Since α is a theorem in $\mathcal{S}_m^*(\mathfrak{A})$ and $\mathcal{I}_0$ is a model of $\mathcal{S}_m^{*0}(\mathfrak{A})$, by Theorem 8, $\alpha_{\mathcal{I}_0} = \vee$ identically. Since free individual variables do not occur in α, $\alpha_{\mathcal{I}_0}$ has the constant value $\vee$ in $\boldsymbol{P}/\nabla$, and it follows from the form of α that

$$\alpha_{\mathcal{I}_0} = \bigcup_{\sigma_1 \in T} \bigcap_{\sigma_2 \in T} \bigcup_{\sigma_3 \in T} \bigcap_{\sigma_4 \in T} \bigcup_{\sigma_5 \in T} [\| \beta(\sigma_1, \sigma_2, \sigma_3, \sigma_4, \sigma_5) \|_{\mathfrak{A}}]$$

where $\bigcup$ and $\bigcap$ denote infinite joins and meets, respectively, in the m-element Post algebra $\boldsymbol{P}/\nabla$. It follows easily therefrom that there are terms τ_i, τ_j, τ_h such that $[\| \beta(\tau_i, x_{k(i)}, \tau_j, x_{p(i,j)}, \tau_h) \|_{\mathfrak{a}}] = \vee$, which contradicts condition (22).

Open theories have various special properties which are not attributes of other theories, as testified, e.g., by Theorem 15 and Theorem 16. It turns out, however, that the difference between theories and open ones is not essential in the sense that any theory $\mathcal{S}_m^*(\mathfrak{a})$ can be transformed into an open one by the adjunction to its language of new functors and by the replacement of the specific axioms by open formulas which, to put it imprecisely, carry the same mathematical meaning. Without any restriction as to generality, it may be assumed that the specific axioms of $\mathcal{S}_m^*(\mathfrak{a})$ are closed prenex formulas. Should it be otherwise, we can pass to a theory $\mathcal{S}_m^*(\mathfrak{a}')$, whose set of theorems is the same as in the case of $\mathcal{S}_m^*(\mathfrak{a})$ and whose specific axioms have the property mentioned above. Next, by using the Skolem method, we eliminate quantifiers from each axiom $\alpha \in \mathfrak{a}$. Roughly speaking, the procedure is as follows. If α is a prenex closed formula in the form

$$\bigcap\xi_1 \bigcup\eta_1 \bigcup\eta_2 \bigcap\xi_2 \bigcup\eta_3\, \beta(\xi_1, \eta_1, \eta_2, \xi_2, \eta_3),$$

then we adjoin to the language of the theory under consideration two new one-argument functors $\phi_\alpha', \phi_\alpha''$, and one two-argument functor ϕ_α''', and instead of the axiom α we adopt a new axiom α^* in the form $\beta(x_1, \phi_\alpha'(x_1), \phi_\alpha''(x_1), x_2, \phi_\alpha'''(x_1, x_2))$, where x_1, x_2 are any two distinct free individual variables. By proceeding similarly in the case of each specific axiom and by adding each time new functors to the language we obtain a language $\mathcal{L}_m^*$, augmented by the functors adjoined as above, and a set $\mathfrak{a}^*$ of formulas obtained from the formulas in $\mathfrak{a}$ by the elimination of quantifiers by the method described above. The theory $\mathcal{S}_m^*(\mathfrak{a}^*)$, with its language $\mathcal{L}_m^*$ and the set $\mathfrak{a}^*$ of specific axioms, called the Skolem extension of $\mathcal{S}_m^*(\mathfrak{a})$, is a non-essential extension of $\mathcal{S}_m^*(\mathfrak{a})$. This means that every formula β in $\mathcal{S}_m^*(\mathfrak{a})$ is a theorem in that theory iff it is a theorem in $\mathcal{S}_m^*(\mathfrak{a}^*)$; hence the sets of theorems in both theories are identical as to the formulas of the initial theory. This is the basic meaning of the theorem on the elimination of quantifiers from the axioms of a theory which holds for the classical theories (the second ϵ-theorem,

Hilbert and Bernays [14]) and can be generalized so as to cover the theories based on m-valued logics in the following form:

THEOREM 17: *For every formula α in a theory $\mathcal{S}_m^*(\mathfrak{A})$, whose specific axioms are closed prenex formulas, α is a theorem in $\mathcal{S}_m^*(\mathfrak{A})$ iff α is a theorem in an open theory $\mathcal{S}_m^*(\mathfrak{A}^*)$ which is the Skolem extension of $\mathcal{S}_m^*(\mathfrak{A})$.*

The proof of Theorem 17 due to E. Perkowska is analogous to that of case of the classical theories (Rasiowa [35]) and is based on the lemma stating that every m-valued model of $\mathcal{S}_m^*(\mathfrak{A})$ can be extended into an m-valued model of $\mathcal{S}_m^*(\mathfrak{A}^*)$.

It is worth mentioning that the Craig interpolation theorem [5] for the classical predicate calculi holds for the predicate calculi based on m-valued logics, $m \geqq 2$. It has the following form. Assume that the formalized language $\mathcal{L}_m^*$ of $\mathcal{S}_m^*$ is functor-free.

THEOREM 18: *Let α and β be formulas of $\mathcal{S}_m^*$, such that $\alpha \vdash \beta$. Then $\alpha \vdash \gamma$ and $\gamma \vdash \beta$ for some formula γ which contains only those predicates that occur in both α and β.*

The proof refers to the case $m = 2$, i.e., to the Craig interpolation theorem (Rasiowa [39]).

We now revert to linear intuitionistic logic. When we pass from the propositional calculus $\mathcal{S}_{\chi l}$ of the linear intuitionistic logic as described in Sec. 1 to a predicate calculus by using the procedure described in Sec. 3, and add to the logical axioms all formulas in the form $(\bigcap\xi(\alpha \cup \beta(\xi)) \Rightarrow (\alpha \cup \bigcap\xi\beta(\xi)))$ [19], where ξ does not occur in α, then we obtain the predicate calculus $\mathcal{S}_{\chi l}^*$, in which Horn [15] has proved the following

THEOREM 19: *For every formula α in $\mathcal{S}_{\chi l}^*$ the following conditions are equivalent:*

[19] Horn [15] proved that they do not follow from the logical axioms obtained from those of $\mathcal{S}_{\chi l}$ by the procedure described in Sec. 3, but satisfy conditions (i)–(v) of Theorem 19 formulated below.

(i) $\vdash \alpha$,

(ii) $\alpha_{\mathfrak{g}} = \vee$ *identically for every interpretation* $\mathfrak{g}$ *in any set* $J \neq 0$ *and in every linear pseudo-Boolean algebra being a chain,*

(iii) $\alpha_{\mathfrak{g}} = \vee$ *identically for every interpretation* $\mathfrak{g}$ *in any set* $J \neq 0$ *and in the linear pseudo-Boolean algebra* R *which is the chain of real numbers with* $\wedge$ *and* $\vee$ *adjoined as least and greatest elements respectively.*

(iv) $\alpha_{\mathfrak{g}} = \vee$ *identically for every interpretation* $\mathfrak{g}$ *in any set* $J \neq 0$ *and in the linear pseudo-Boolean algebra* C *which is the chain of rational numbers with* $\wedge$ *and* $\vee$ *adjoined as least and greatest elements respectively,*

(v) $\alpha_{\mathfrak{g}} = \vee$ *identically for every interpretation* $\mathfrak{g}$ *in any set* $J \neq 0$ *and in every linear pseudo-Boolean algebra which is an enumerable chain.*

As a corollary from Theorem 19 and Theorem 11 we obtain the following

THEOREM 20: *If a formula* α *of the predicate calculus* $\mathcal{S}^*_{\chi l}$ *is a theorem of* $\mathcal{S}^*_{\chi l}$, *then it is a theorem in* $\mathcal{S}^*_m$ *for every* $m = 2, 3, \cdots$.

The converse statement, contrary to the case of the corresponding propositional calculi (see Theorem 6 in Sec. 2), does not hold. For instance the formula $(\neg \bigcap \xi \rho(\xi) \Rightarrow \bigcup \xi \neg \rho(\xi))$ is a theorem in each m-valued predicate calculus $\mathcal{S}^*_m$, but it is not a theorem in $\mathcal{S}^*_{\chi l}$, since it does not satisfy the condition (iii) in Theorem 19. To prove this, assume that J is the set of positive integers and that $(w_j)_{j \in J}$ is a sequence of all rational numbers. Put $\rho_{\mathfrak{g}}(j) = w_j$, $j \in J$. Then $(\neg \bigcap \xi \rho(\xi))_{\mathfrak{g}} = \vee$ and $(\bigcup \xi \neg \rho(\xi))_{\mathfrak{g}} = \wedge$.

Further research now being conducted on those many-valued logics which have algebraic analogues in classes of certain Post algebras follows the direction indicated by the advances in the theory of Post algebras, namely that of generalizations of these algebras, and hence also, generalizations of the logical systems related with the former.

BIBLIOGRAPHY

1. Białynicki-Birula, A. and H. Rasiowa, "On constructible falsity in the constructive logic with strong negation", *Colloq. Math.*, **6** (1958), 287–310.
2. Birkhoff, G., "On the structure of abstract algebras," *Proc. Cambridge Philos. Soc.*, **31** (1935), 433–454.
3. Cat Ho, N., "Generalized Post algebras and their applications to some infinity many-valued logics", Doctoral thesis, University of Warsaw, 1971, *Dissertationes Math.*, **107** (1973).
4. Chang, C. C. and A. Horn, "Prime ideals characterization of generalized Post algebras", Lattice theory, *Proceedings of Symposium in Pure Mathematics*, **2** (1961), 43–48.
5. Craig, W., "Linear reasoning—A new form of the Herbrand-Gentzen theorem", *J. Symbolic Logic*, **22** (1957), 250–268.
6. Dummet, M., "A propositional calculus with denumerable matrix", *J. Symbolic Logic*, **24** (1959), 97–106.
7. Dwinger, P., "Notes on Post algebras", I, II, *Indag. Math.*, **28** (1966), 462–478.
8. ———, "Generalized Post algebras", *Bull. Acad. Polon. Sci. Sér. Sci. Math. Astronom. Phys.*, **16** (1958), 559–563.
9. ———, "Ideals in generalized Post algebras", ibidem, **17** (1969), 483–486.
10. Epstein, G., "The lattice theory of Post algebras", *Trans. Amer. Math. Soc.*, **95** (1960), 300–317.
11. Henkin, L., "An algebraic characterization of quantifiers", *Fund. Math.*, **37** (1950), 63–74.
12. Heyting, A., "Die formalen Regeln der intuitionistischen Logik", *Sitzungsberichte der Preussischen Akademic der Wissenschaften, Phys. Mathem. Klasse* (1930), 42–56.
13. ———, "Intuitionism, an Introduction", *Studies in Logic and Foundations of Mathematics*, Amsterdam: North Holland, 1956.
14. Hilbert, D. and P. Bernays, *Grundlagen der Mathematik*, vol. I, Berlin: Springer, 1934 (repr. Ann Arbor, Michigan, 1944), vol. II, Berlin: Springer, 1939 (repr. Ann Arbor, Michigan, 1944).
15. Horn, A., "Logic with truth values in a linearly ordered Heyting algebra", *J. Symbolic Logic*, **34** (1969), 395–408.
16. Johansson, I., "Der Minimalkalkül, ein reduzierter intuitionistischer Formalismus", *Compositio Math.*, **4** (1936).

17. Lewis, C. I. and C. H. Langford, "Symbolic Logic", New York: 1932.

18. Łoś, J., "Algebraic treatment of the methodology of elementary deductive systems", *Studia Logica*, **2** (1955), 151–212.

19. ———, "Quelques remarques, théorèmes et problèmes sur les classes définissables d'algèbres", *Mathematical Interpretation of Formal Systems, Studies in Logic and Foundations of Mathematics*, North Holland, 1955, pp. 98–113.

20. Łukasiewicz, J., "O logice trójwartościowej", *Ruch filozoficzny*, **5** (1920), 169–170.

21. ———, "Elementy logiki matematycznej", *Warszawa*, 1929 (English trans. "Elements of mathematical logic", Oxford 1963, International Series of Monographs on Pure and Applied Mathematics, Vol. 31).

22. Łukasiewicz, J. and A. Tarski, "Untersuchunegen über den Aussagenkakül", *Comptes rendus des séances de la Société des Sciences et des Lettres de Varsovie*, Cl. III, **23** (1930), 30–50.

23. MacNeille, H., "Partially ordered sets", *Trans. Amer. Math. Soc.*, **42** (1937), 416–460.

24. Malcev, A., "Iterativnije algebri i mnogoobrazija Posta", *Algebra i Logika*, **5** (1966), 8–24.

25. Markov, A. A., "Konstruktivnaja logika", *Uspehi Mat. Nauk.*, **5** (1950), 187–188.

26. McKinsey, J. C. C. and A. Tarski, "The algebra of topology", *Ann. of Math.*, **45** (1944), 141–191.

27. ———, "On closed elements in closure algebras", *ibidem*, **47** (1946), 122–162.

28. ———, "Some theorems about the sentential calculi of Lewis and Heyting", *J. Symbolic Logic*, **13** (1948), 1–15.

29. Moisil, G. C., "Sur les logiques de Łukasiewicz à un nombre fini de valeurs", *Rev. Roumaine Math. Pures Appl.*, **9** (1964), 905–920.

30. Mostowski, A., "Proofs of non-deducibility in intuitionistic functional calculus", *J. Symbolic Logic*, **13** (1948), 204–207.

31. Nelson, D., "Constructible falsity", *J. Symbolic Logic*, **14** (1949), 16–26.

32. Post, E. L., "Introduction to a general theory of elementary propositions", *Amer. J. Math.*, **43** (1921), 165–185.

33. Rasiowa, H., "Algebraic treatment of the functional calculi of Lewis and Heyting", *Fund. Math.*, **38** (1951), 99–126.

34. ———, "Algebraic models of axiomatic theories", *Fund. Math.*, **41** II (1954), 291–310.

35. ———, "On ϵ-theorems", *Fund. Math.*, **43** II (1956), 156–164. Errata, *ibidem*, **44** III (1957), 333.

36. ———, "N-lattices and constructive logic with strong negation", *ibidem*, **46** (1958), 61–80.

37. ———, "A theorem on the existence of prime filters in Post algebras and the completeness theorem for some many-valued predicate calculi", *Bull. Acad. Polon. Sci. Sér. Sci. Math. Astronom. Phys.*, **17** (1969), 347–354.

38. ———, "Ultraproducts of *m*-valued models and a generalization of the Löwenheim-Skolem-Gödel-Malcev theorem for theories based on *m*-valued logics", *ibidem*, **18** (1970), 415–420.

39. ———, "The Craig interpolation theorem for m-valued predicate calculi", *ibidem*, **20** (1972), 341–346.

40. Rasiowa, H. and R. Sikorski, "Algebraic treatment of the notion of satisfiability", *Fund. Math.*, **40** (1953), 62–95.

41. ———, "The Mathematics of Metamathematics", *Monografie Matematyczne*, **41**, Warszawa, 1963 (3rd edition 1970).

42. Rosenbloom, P. C., "Post algebras I. Postulates and general theory", *Amer. J. Math.*, **64** (1942), 167–188.

43. Rosser, J. B., *Simplified Independence Proofs*, New York and London: Academic Press, 1969.

44. Rosser, J. B. and A. R. Turquette, "Many-valued logics", *Studies in Logic and Foundations of Mathematics*, Amsterdam: North-Holland, 1952.

45. Rousseau, G., "Logical systems with finitely many truth-values", *Bull. Acad. Polon. Sci., Sér. Sci. Math., Astronom. Phys.*, **17** (1969), 189–194.

46. ———, "Post algebras and pseudo-Post algebras", *Fund. Math.*, **67** (1970), 133–145.

47. Sawicka, H., "On some properties of Post algebras with countable chain of constants", *Colloq. Math.*, **25** (1972), 201–209.

48. Sikorski, R., "Products of abstract algebras", *Fund. Math.*, **39** (1952), 211–228.

49. ———, "Some applications of interior mappings", *Fund. Math.*, **45** (1958), 200–212.

50. Stone, M. H., "The theory of representation for Boolean algebras", *Trans. Amer. Math. Soc.*, **40** (1936), 37–111.

51. ———, "Topological representation of distributive lattices and Brouwerian logics", *Casopis Pěst. Mat. Fysiky*, **67** (1937), 1–25.

52. Tarski, A., "Der Aussagenkalkul und die Topologie", *Fund. Math.*, **31** (1938), 103–134.

53. Traczyk, T., "On axioms and some properties of Post algebras", *Bull. Acad. Polon. Sci., Sér. Sci. Math. Astronom. Phys.*, **10** (1962), 509–512.

54. ———, "Some theorems on independence in Post algebras", *ibidem*, **11** (1963), 3–8.

55. ———, "Axioms and some properties of Post algebras", *Colloq. Math.*, **10** (1963), 198–209.

56. ———, "A generalization of Loomis-Sikorski's theorem", *ibidem*, **12** (1964), 155–161.

57. ———, "An equational definition of a class of Post algebras", *Bull. Acad. Polon. Sci., Sér. Sci. Math. Astronom. Phys.*, **12** (1964), 147–149.

58. ———, "Weak isomorphism of Boolean and Post algebras", *Colloq. Math.*, **13** (1965), 159–194.

59. ———, "On Post algebras with uncountable chain of constants", *Bull. Acad. Polon. Sci. Sér. Sci. Math. Astronom. Phys.*, **15** (1967), 673–680.

60. ———, "Prime ideals in generalized Post algebras", *ibidem*, **16** (1968), 369–373.

61. Włodarska, E., "On the representation of Post algebras preserving some infinite joins and meets", *ibidem*, **18** (1970), 49–54.

62. Willis, V. J., "Polyadic Post algebras", Doctoral Thesis, the University of Alberta (Canada), 1967.

FROM SHEAVES TO LOGIC

Gonzalo E. Reyes(1)

INTRODUCTION

1. This paper is concerned with the categorical approach to logic started by F. W. Lawvere [L1, L2] and further developed by Lawvere himself, H. Volger [V1, V2] and A. Joyal [J1, J2].

To motivate and describe the scope of this approach, we shall distinguish the following aspects of an elementary theory (say the elementary theory of integral domains of characteristic 0):

(1) SYNTACTICAL OR ARITHMETICAL

The theory in question is formulated in English enriched with some mathematical symbols: $+$, $\cdot$, 0, 1, etc. By idealization, we

(1) The first draft of this paper was written while in residence at the Sherbrooke branch of the 1972 Summer Research Institute of the Canadian Mathematical Congress.

view the formulas of this language as finite strings of symbols of a fixed vocabulary. The grammar of this language is assumed to be describable by purely mechanical rules. Furthermore, the theorems themselves are identified with certain formulas also generated by mechanical rules from a given subset of strings called the axioms. In our case the axioms are strings such as $\forall x \forall y \forall z ((x + y) + z = x + (y + z))$, $1 + 1 \neq 0$, etc.

(2) POLYADIC OR ALGEBRAIC

Here we are concerned with the concepts or notions expressible in the theory: "unit," "solution of the equation $p(X) = 0$," "double" (or equal to $2y$ for some y), "quadruple," etc. These concepts are identified with equivalence classes of formulas, two formulas being equivalent when they are provably equivalent in the theory. For instance, "unit" is the equivalent class of the formula $\exists y (xy = 1)$. These equivalence classes can be made into algebraic structures (cylindric, polyadic algebras) which are then studied in the spirit of lattice theory and general algebra. Traditionally, these algebras have been the subject matter of algebraic logic.

(3) CONCEPTUAL OR ALGEBRO-GEOMETRIC

As in (2), we are concerned with the concepts that can be expressed in the theory. However, we are also interested in the relations of a given concept to the rest and in ways of going from one concept to another. For instance it is clear that the concept of "solution of the equation $p(X) \cdot q(Y) = 0$" is in some sense the cartesian product of the concepts "solution of the equation $p(X) = 0$" and "solution of the equation $q(X) = 0$." Again, we can go from "quadruple" to "double" in several ways, for instance by "halving." Technically, we are interested in a *category* of concepts rather than in algebras. The objects of the category will be equivalence classes of formulas. Roughly, two formulas are equivalent if they are provably equivalent in the theory. The morphisms will be equivalence classes of definable relations provably functional in the theory.[2] These categories are then studied in the spirit of category theory and more specifically of algebraic geometry, as developed for instance by the Grothendieck School.

(4) SEMANTICAL OR SET-THEORETICAL

This is the aspect concerned with the meaning of the symbols and the truth or validity of the sentences of the theory. Traditionally, these notions have been defined almost exclusively in terms of set-theoretical realizations or models of the theory. Thus, in our case, a sentence is valid if it holds in all integral domains of characteristic 0. Following the lead of algebraic geometry, we also consider realizations of the theory in other set-like categories such as topoi, for instance. From an operational point of view, the nice things about realizations is the existence of good limits and good exactness conditions. Furthermore, some perfectly consistent and respectable theories (intuitionistic, infinitary) do not possess set-theoretical realizations though they have models in topoi. The work of Lawvere-Tierney on elementary topoi has also shown that Cohen-type independence results in set theory may be viewed as constructions of models in topoi for axiomatic set theory. From this point of view, it seems more natural to subsume (4) under (3).

It is clear from this description that the "Boolean" categories obtained from elementary theories are just one of a series. Indeed, intuitionistic theories give rise to "Heyting" categories. Moreover, other "conceptual" categories can be obtained by selecting only some of the logical operations in either the classical or intuitionistic case.

One technical advantage of these categories is the existence of finite $\varprojlim$. In this new context, Joyal gave an elegant, conceptual proof of the completeness theorem for "Boolean" categories. More recently he has used his method to give unified proofs of the completeness theorems of Gödel and Kripke, as well as the em-

(2) The details of this construction will appear in the Master's thesis of M. Jean Dionne at the Université de Montréal.

bedding theorems of Mitchell and Barr [B], for abelian and regular categories, respectively.

Interestingly enough, these "conceptual" categories had already appeared in topology. Every topological space, for instance, gives rise to a category of set-valued sheaves (or "espaces étalés") which is a category of this type. In particular, logical operations $\forall$, $\exists$, $\Rightarrow$, $\wedge$, $\vee$ appear as operations on these topological spaces (i.e., "espaces étalés") and continuous functions rather than on propositional functions. Thus any topological space has, so to speak, an internal logic, which is not Boolean, in general.

As an example, let us consider the circle Π with its natural topology. The category of "espaces étalés" over Π is defined as follows: an *object* is a topological space E together with a continuous projection p over Π which is a local homeomorphism (i.e., $\forall x \in E$ has an open neighborhood U such that $p \mid U$ is a homeomorphism onto an open neighborhood of $p(x) \in \Pi$). A typical object is the helix

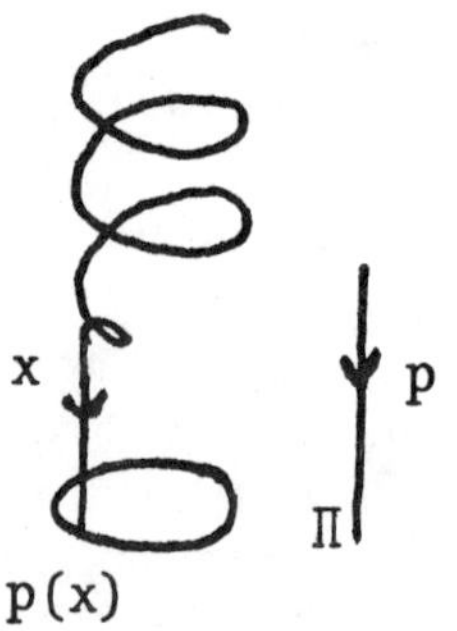

The morphisms are continuous functions $E \xrightarrow{f} E'$ such that the diagram

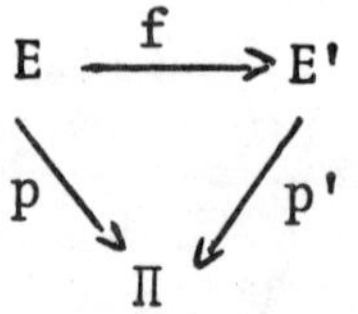

commutes.

The "logic of the circle" can be considered to be a many sorted predicate calculus (one sort for each "espace étalé"). It may be obtained with the help of the following dictionary:

sentence	:	open subset of Π
sort (of variable x)	:	"espace étalé" X over Π
formula $\varphi(x, y)$	:	"sous espace étalé" of the fibered product $X \underset{\Pi}{\times} Y$
conjunction (with same variables)	:	set-theoretical intersection
disjunction (with same variables)	:	"espace étalé" associated with the union
conjunction (with disjoint variables)	:	fiber product over Π
negation (of a sentence)	:	largest open set of Π disjoint from a given open set
existential quantifier $\exists x\varphi(x, y)$	:	"espace étalé" associated with the image of the corresponding subspace of $X \underset{\Pi}{\times} Y$ under the second projection $X \underset{\Pi}{\times} Y \xrightarrow{\Pi^2} Y$
constant	:	global section $s: \Pi \xrightarrow{s} E$, where E is "étalé" over Π
theorem	:	Π

This (partial) dictionary suffices to see that the "logic of the circle" is not Boolean. In fact if σ is the "sentence" $\sigma = \{\theta \in \Pi: 0 < \theta < \pi) \cup \{\theta \in \Pi: \pi < \theta < 2\pi\}$, then $\neg\neg\sigma = \Pi$. We notice, however, that this logic is intuitionistic (II.2.6).

More important is the fact that we could also study "the arithmetic of the circle," "the analysis of the circle" and the "set-theory of the circle" in a similar fashion. For the arithmetic, for instance, we could extend our dictionary: set of natural numbers: "espace étalé" $N \times \Pi$, etc., etc.

In this paper, however, we shall concentrate on the "predicate calculus" of these geometric constructions.

The work of Grothendieck, Giraud, Verdier [V] showed that the construction of set-valued sheaves could be carried through, starting from *any* small category and not only from a topological space. These logical, arithmetical and analytic aspects of these geometric constructions were discovered and emphasized by Lawvere-Tierney [KW], [L3] who also showed that the external notion of infinite limit of the Grothendieck topos could be eliminated. The fundamental notion of *elementary topos* unifying the logical and geometric aspects of these constructions was thus isolated.

2. We divide our work into four sections. The first consists of the definitions and elementary theory of regular, Heyting and Boolean categories. It was practically written by Joyal in an abortive scheme to publish a joint paper.

The second is expository and is my own doing. The notions of sites and sheaves are defined and properties used in the sequel are stated without proofs.

The third deals with the problem of unifying the geometric and logical aspects of a Grothendieck topos. Only partial and "compromise" solutions can be expected, since the external notion of infinite limit precludes a satisfactory solution (§2). Boolean valued sets appear naturally as an instance of the "compromise solution." This section started from an observation of Joyal (Theorem III 1.7). Although several of the positive results are easy consequences of the general theory, the counter-examples seem new.

Finally, the first two sections of the last chapter are based on lectures given by Joyal at the Université du Québec à Montréal and the "Atelier d'Expression Tautologique" at the Université de Montréal. The last section is based on lectures by myself at the Université de Montréal. However, it mostly belongs to the Montreal logical folklore.

There are no logical prerequisites to read this paper. As far as category theory is concerned, we shall assume familiarity with the material presented in the first two chapters of Pareigis [P]. This paper is one of a series of studies devoted to develop the algebrogeometric aspect of logic. The aggressive content of this program is to "geometrize" logic in a sense that should become clearer to the readers (and hopefully to the author or authors) as we proceed.

Our debt of gratitude to André Joyal is obvious from the preceding. It is equally obvious, however, that he should not be made responsible for all of what is here said and done. Lately, we had the help of Hugo Volger on several categorical matters. Finally, Jean Dionne and D. A. Higgs corrected some mistakes in a first draft. Many thanks to them.

SECTION I

CONCEPTUAL CATEGORIES

1. REGULAR CATEGORIES

Regular categories are locally small categories which are rather poor in logical operations. Indeed, they reflect the operations of substitution, conjunction and existential quantifiers only:

DEFINITION 1.1: A category $\mathcal{A}$ is *regular* if

(i) Finite left limits ($\lim_{\leftarrow}$) exist in $\mathcal{A}$;

(ii) For every $(A \xrightarrow{f} B) \in \mathcal{A}$ there is a smallest sub-object $I \rightarrowtail B$ through which f factors, i.e.,

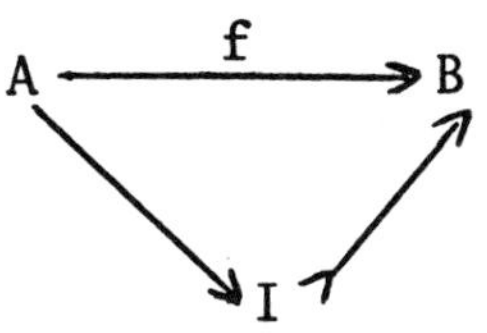

We shall write $I = Im(f)$.

(iii) For every pull-back square

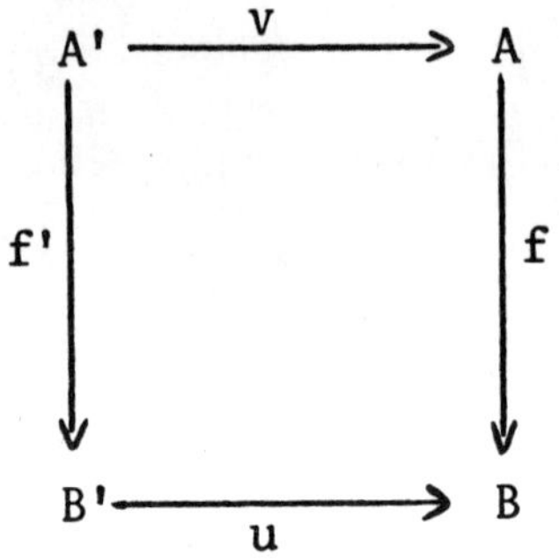

if $Im(f) = B$, then $Im(f') = B'$.

Remark 1.2: Condition (i) of the previous definition is equivalent to the conjunction of the following three conditions:

C1. There is a final object $1 \in |\mathcal{A}|$.
C2. The product $A \times B$ of two objects $A, B \in |\mathcal{A}|$ exists.
C3. The intersection $I \wedge I'$ of two sub-objects $I \rightarrowtail A$ $I' \rightarrowtail A$ always exists.

Notation 1.3:

(1) We shall write $\mathcal{P}(A)$ for the set of all sub-objects of A.

(2) For every $(A \xrightarrow{f} B) \in \mathcal{A}$ we shall write $f^*: \mathcal{P}(B) \rightarrow \mathcal{P}(A)$ to denote the function which associates with every sub-object $I \rightarrowtail B$ the sub-object $J \rightarrowtail A$ such that the square

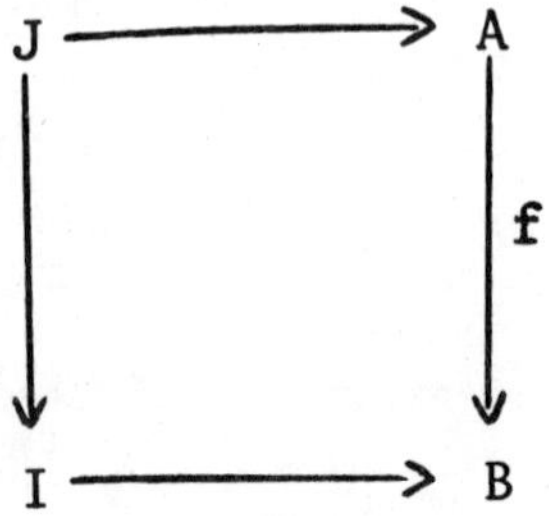

is a pull-back.

(3) For every $(A \xrightarrow{f} B) \in \mathcal{A}$ and every $I \overset{j}{\rightarrowtail} A$, we shall denote by $f(I)$ or by $\exists_f(I)$ the sub-object $Im(f \circ j)$ of B.

(4) We shall use the notation $A \overset{f}{\twoheadrightarrow} B$ to indicate that $Im(f) = B$.

With these notations, condition (iii) of definition 1.1 can be replaced by the conjunction of the following two conditions:

I(1) For every $(A \xrightarrow{f} B) \in \mathcal{A}$ and every $C \in |\mathcal{A}|$ if $Im(f) = B$, then $Im(1_C \times f) = C \times B$.

I(2) For every $(A \xrightarrow{f} B) \in \mathcal{A}$ and for every $I \rightarrowtail B$ if $Im(f) = B$, then there is $J \rightarrowtail A$ such that $f(J) = I$.

To show this, we just notice that every morphism $B' \xrightarrow{u} B$ factors into a monomorphism followed by a projection:

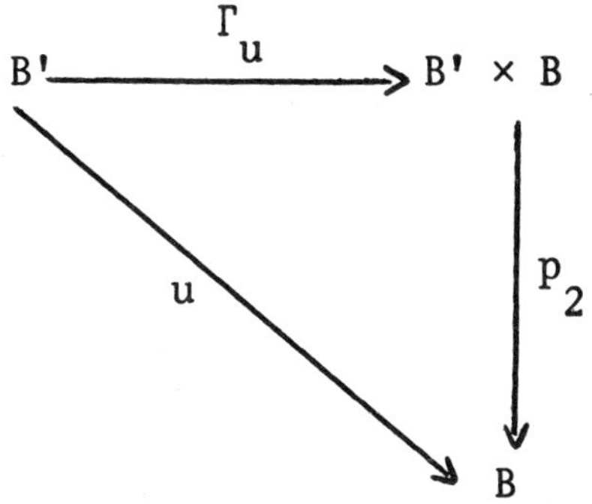

where $\Gamma_u = [1_{B'}, u]$ is the graph of u.

Our condition (iii) of definition 1.1 splits in two: the first for $B' \times B \xrightarrow{p_2} B$ and the second for $B' \overset{\Gamma_u}{\rightarrowtail} B' \times B$.

Remark 1.4: $\mathcal{P}$ is a contravariant functor from the category $\mathcal{A}$ to the category of $\wedge$-semilattices with greatest element, i.e., $(fg)^* = g^*f^*$ and $f^*(I \wedge I') = f^*(I) \wedge f^*(I')$. On the other hand, $f^*: \mathcal{P}(B) \to \mathcal{P}(A)$ is a functor between the categories $\mathcal{P}(A)$, $\mathcal{P}(B)$ (since these sets are ordered and f^* is a monotone increasing function). We obviously have

$$\exists_f(I) \leqq J \quad \text{iff} \quad I \leqq f^*(J)$$

and this shows that the monotone function

$$\exists_f: \mathcal{P}(A) \to \mathcal{P}(B)$$

is left adjoint to f^* ($\exists_f \dashv f^*$).

Since the composition of adjoints is adjoint to the composition of functors, $\exists_f \exists_g = \exists_{fg}$. For the rest of this section, we shall work in a regular category $\mathcal{A}$.

Proposition 1.5: *If $Im(f) = B$, then $A \xrightarrow{f} B$ is an epimorphism.*

Definition 1.6: An epimorphism $A \xrightarrow{f} B$ is *special* if $Im(f) = B$, i.e., if $A \overset{f}{\twoheadrightarrow} B$.

For regular categories this auxiliary notion has the following characterization:

Proposition 1.7: *An epimorphism is special iff it is a regular epimorphism.*

Sufficiency.

Let us consider a factorization

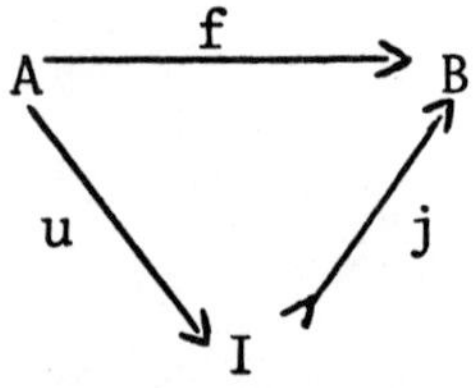

where f is a regular epimorphism. We then have

$$f = \text{Coker}\left(A \underset{B}{\times} A \overset{p_1}{\underset{p_2}{\rightrightarrows}} A\right).$$

Since $fp_1 = fp_2$ and j is a monomorphism, $up_1 = up_2$. This shows that there is a $v: B \to I$ such that $vf = u$. We conclude that $vj = 1_I$ and this implies that j is an isomorphism. Necessity will be proved later. (Theorem 1.21.)

PROPOSITION 1.8: *Every special epimorphism which is also a monomorphism is an isomorphism.*

Proof: Immediate.

PROPOSITION 1.9: *The composition of two special epimorphisms is again special.*

Proof: It suffices to use the formula

$$\exists_f \exists_g = \exists_{fg}.$$

PROPOSITION 1.10: *Assume that in the commutative diagram*

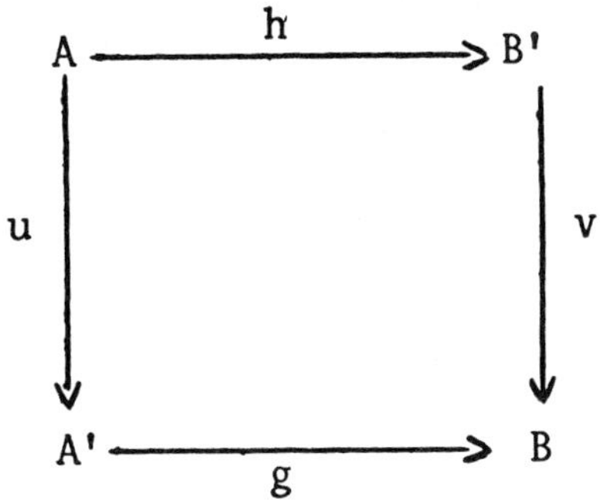

u is a special epimorphism and v is a monomorphism. Then there is exactly one arrow $A' \xrightarrow{f} B'$ such that $fu = h$ and $vf = g$.

Proof: We have $Im(g \circ u) = Im(g)$, since u is a special epimorphism. Similarly, $Im(g \circ u) = Im(vh) \leqq B' \overset{v}{\rightarrowtail} B$, since v is monic. This shows that $Im(g) \leqq B' \overset{v}{\rightarrowtail} B$, i.e., there exists $f: A' \to B'$ such that

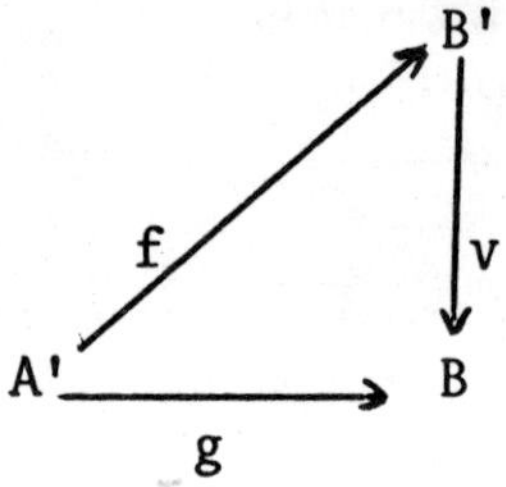

is commutative. We obtain $vh = gu = vfu$ and this implies that $h = fu$, since v is a monomorphism.

COROLLARY 1.11: *If*

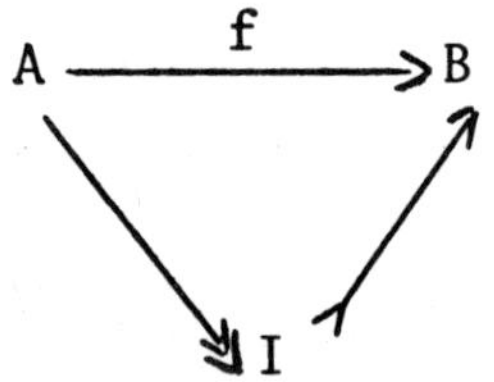

is a factorization of f into a special epimorphism and a monomorphism, then $I = Im(f)$.

PROPOSITION 1.12: *Let*

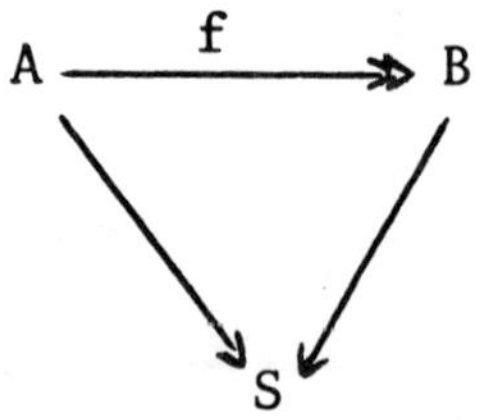

be a special epimorphism above S. Then for every $S' \to S$, the morphism

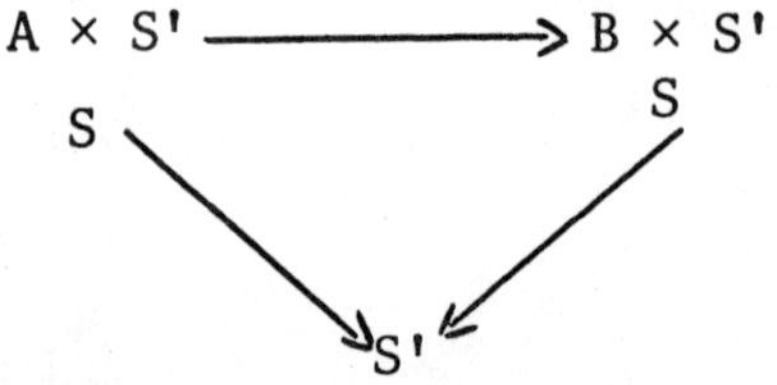

above S', obtained by pulling back f along $S' \to S$, is special.

Proof: It suffices to use the appropriate converse of the fact that the composition of two pull-backs is again a pull-back.

Proposition 1.13: *If the diagram*

$$\begin{array}{ccc} A & \xrightarrow{f} & B \\ {\scriptstyle u}\downarrow & & \downarrow{\scriptstyle v} \\ A' & \xrightarrow[f']{} & B' \end{array}$$

is a pull-back, then the square

$$\begin{array}{ccc} P(A) & \xrightarrow{\exists_f} & P(B) \\ {\scriptstyle u^*}\uparrow & & \uparrow{\scriptstyle v^*} \\ P(A') & \xrightarrow[\exists_{f'}]{} & P(B') \end{array}$$

is commutative.

Proof: Let $I \overset{i'}{\rightarrowtail} A'$. Then we have a pull-back square

$$\begin{array}{ccc} u^*(I) & \overset{i}{\rightarrowtail} & A \\ \downarrow & & \downarrow{\scriptstyle u} \\ I & \underset{i'}{\rightarrowtail} & A' \end{array}$$

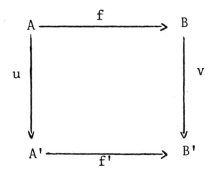

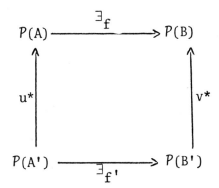

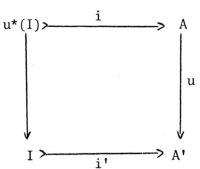

Since the composition of two pull-backs is again a pull-back we obtain a new pull-back square

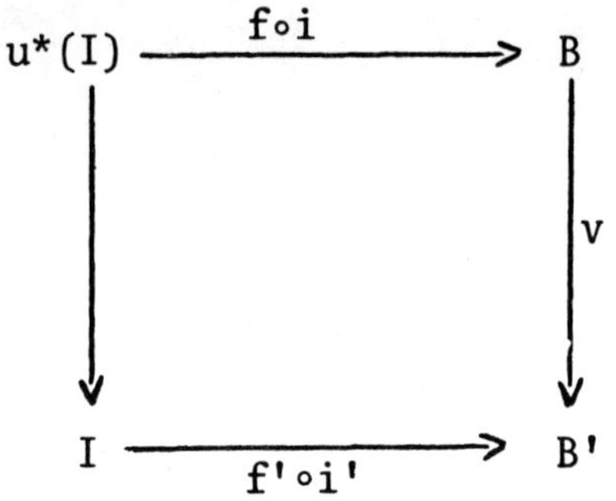

and hence a commutative diagram

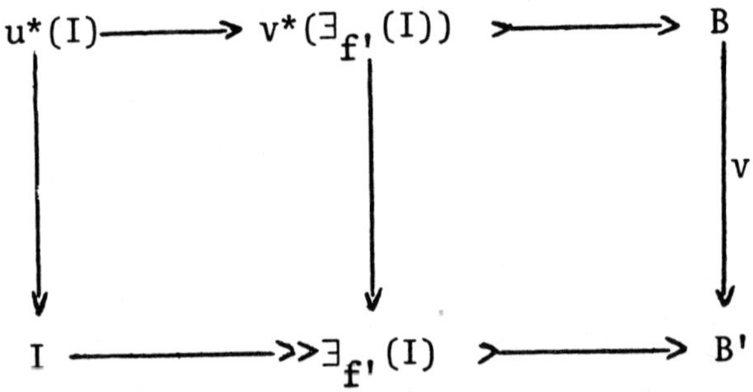

such that the left square is a pull-back. This shows that $u^*(I) \to v^*(\exists_{f'}(I))$ is a special epimorphism just as $I \to \exists_{f'}(I)$. We conclude that

$$\exists_f(u^*(I)) = Im(f \circ i) = v^*(\exists_{f'}(I)).$$

PROPOSITION 1.14: *For every* $A \xrightarrow{f} B$ *and every* $I \in \mathcal{P}(A)$, $J \in \mathcal{P}(B)$

$$\exists_f(f^*(J) \wedge I) = J \wedge \exists_f(I).$$

Proof: It suffices to apply the previous proposition to the pull-back square

$$\begin{array}{ccc} f^*(J) & \longrightarrow & J \\ \downarrow & & \downarrow \\ A & \longrightarrow & B \end{array}$$

COROLLARY 1.15: *For every* $A \xrightarrow{f} B$ *and every* $J \in \mathcal{P}(B)$

$$\exists_f(f^*(J)) = J \wedge Im(f).$$

COROLLARY 1.16: *For every special epimorphism* $A \overset{f}{\twoheadrightarrow} B$ *and every* $J \in \mathcal{P}(B)$

$$\exists_f(f^*(J)) = J.$$

In particular, $f^*\colon \mathcal{P}(B) \to \mathcal{P}(A)$ *is injective.*

PROPOSITION 1.17: *If* $A \overset{f}{\twoheadrightarrow} B$ *and* $C \overset{g}{\twoheadrightarrow} D$, *then* $A \times C \overset{f\times g}{\twoheadrightarrow} B \times D$.

Proof: Since $f \times g = (f \times 1_D) \circ (1_A \times g)$, it is sufficient to use proposition 1.9.

COROLLARY 1.18: $Im(f \times g) = Im(f) \times Im(g)$

PROPOSITION 1.19: *For every* $A \xrightarrow{f} B \xrightarrow{g} C$ $(f \times f)^*(R_g) = R_{gf}$, *where* R_g *and* R_{gf} *are the kernel pairs of* g *and* gf, *respectively.*

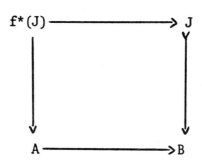

Proof: The following square is a pull-back

$$\begin{array}{ccc} A \underset{C}{\times} A & \longrightarrow & B \underset{C}{\times} B \\ \downarrow & & \downarrow \\ A \times A & \xrightarrow[f \times f]{} & B \times B \end{array}$$

PROPOSITION 1.20: *Using the notations of the previous proposition, if f is a special epimorphism and if $R_f = R_{gf}$, then g is a monomorphism.*

Proof: By the previous proposition,

$$(f \times f)^*(R_g) = R_{gf}$$

$$(f \times f)^*(R_{1_B}) = R_{1_B f} = R_f.$$

But $(f \times f)^*\colon \mathcal{P}(B \times B) \to \mathcal{P}(A \times A)$ is injective, since $f \times f$ is a special epimorphism (propositions 1.16, 1.17). We conclude that $R_{1_B} = R_g$ and this means precisely that g is a monomorphism.

THEOREM 1.21: *Every special epimorphism is regular.*

Proof: Assume that $A \overset{p}{\twoheadrightarrow} B$ is a special epimorphism. Let us show that

$$A \underset{B}{\times} A \underset{p_2}{\overset{p_1}{\rightrightarrows}} A \overset{p}{\to} B$$

is right exact. Let $A \overset{f}{\to} C$ be such that $fp_1 = fp_2$. Let us consider

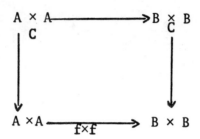

the following diagram

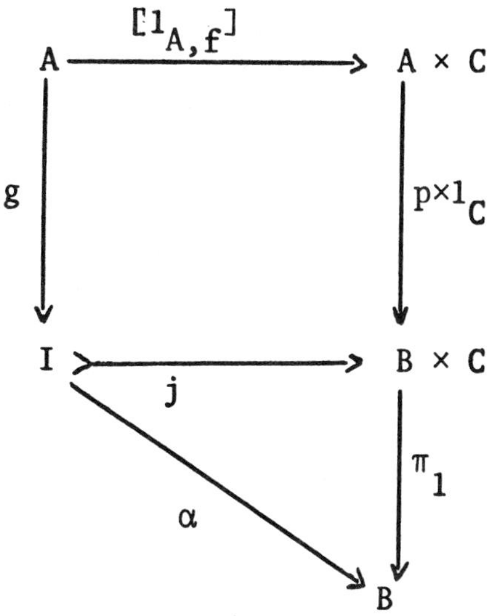

where $A \overset{g}{\twoheadrightarrow} I \overset{j}{\rightarrowtail} B \times C$ is the canonical factorization of $(p \times 1_C) \circ [1_A, f] = [p, f]$ and where $\alpha = \pi_1 \circ j$. We shall prove that α is an isomorphism:

(i) α is a special epimorphism since

$$\begin{aligned} Im(\alpha) &= Im(\alpha \circ g) \\ &= Im(\pi_1 \circ [p, f]) \\ &= Im(p) = B. \end{aligned}$$

(ii) α is a monomorphism. By the previous proposition, it is enough to show that $R_g = R_{\alpha g}$. But $R_g = R_{jg}$, since j is monic and this shows that $R_g = R_{jg} = R_{[p,f]} = R_p \cap R_f$. By hypothesis, $fp_1 = fp_2$ and this means that $R_p \subset R_f$ and hence we obtain $R_g = R_p$. On the other hand $p = \alpha g$. This completes the proof of the fact that α is an isomorphism (proposition 1.8). We thus obtain a morphism $h = \pi_2 j\alpha^{-1}: B \rightarrow C$ rendering the triangle

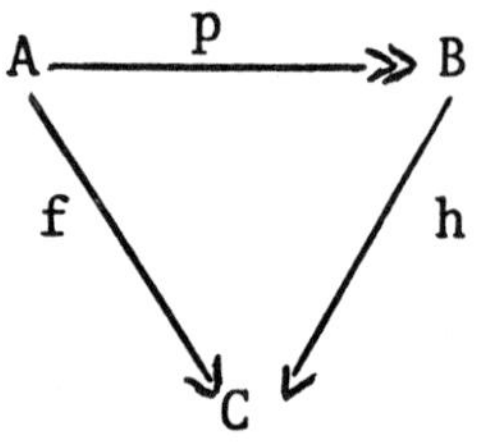

commutative. Q.E.D.

2. ENRICHED REGULAR CATEGORIES

DEFINITION 2.1: Let $f^*: E \rightarrow F$ be an order-preserving function between two partially ordered sets. We say that f^* has a left adjoint $\exists_f: F \rightarrow E$ (right adjoint $\forall_f: F \rightarrow E$) if f^* considered as a functor between the obvious categories has a left (right) adjoint and we write $\exists_f \dashv f^*$ ($f^* \vdash \forall_f$). This is expressed by

$$\forall x \in E \ \forall y \in F \ \ y \leqq f^*(x) \quad \text{iff} \quad \exists_f y \leqq x$$

and

$$\forall x \in E \ \forall y \in F \ \ f^*(x) \leqq y \quad \text{iff} \quad x \leqq \forall_f y,$$

respectively.

PROPOSITION 2.2: *Assume that in the following diagram of order preserving maps*

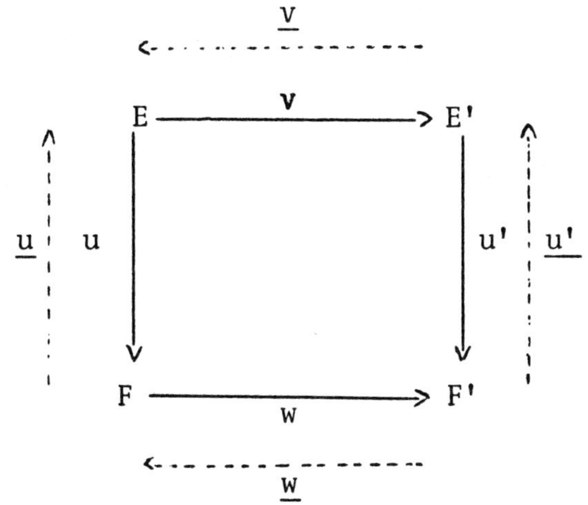

the dotted arrows are right adjoints to the solid ones. Then the dotted diagram is commutative iff the solid diagram is commutative.

COROLLARY 2.3: *The commutativity of each of the following diagrams implies that of the other:*

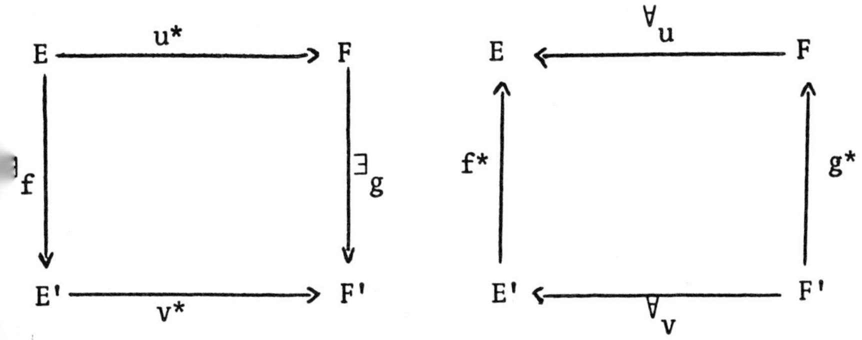

DEFINITION 2.4: An *implicative semilattice* is an $\wedge$-semilattice with maximal element 1 such that for every $x \in F$, the map

$$x \wedge (\ ): F \to F \text{ has a right adjoint,}$$

$$x \Rightarrow (\ ): F \to F.$$

Proposition 2.5: *For an order preserving map* $E \xrightarrow{f^*} F$ *with left adjoint* $\exists_f$ *between two implicative semilattices, the following conditions are equivalent:*

(1) $\forall x, y \in E\ f^*(x \Rightarrow y) = f^*(x) \Rightarrow f^*(y)$,

(2) $\forall x \in E \quad \forall y \in F \quad \exists_f(y) \wedge x = \exists_f(y \wedge f^*(x))$.

Proof: Look at the following diagrams and use Proposition 2.2.

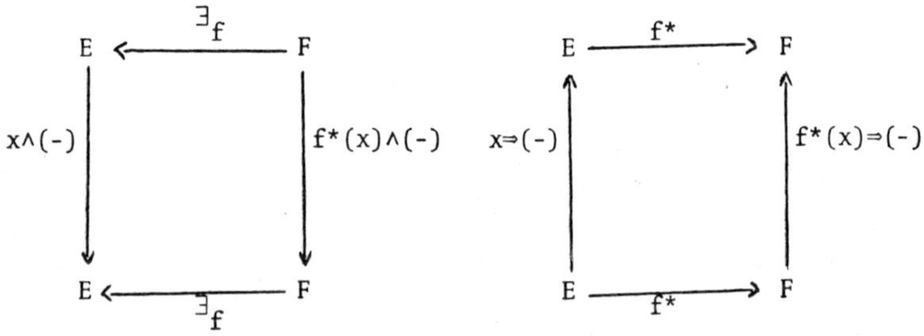

Corollary 2.6: *Assume that* $E \xrightarrow{f^*} F$, $G \xrightarrow{g^*} H$ *are order-preserving maps of implicative semi-lattices having left adjoints* $\exists_f$, $\exists_g$ *respectively. Assume furthermore that the diagram*

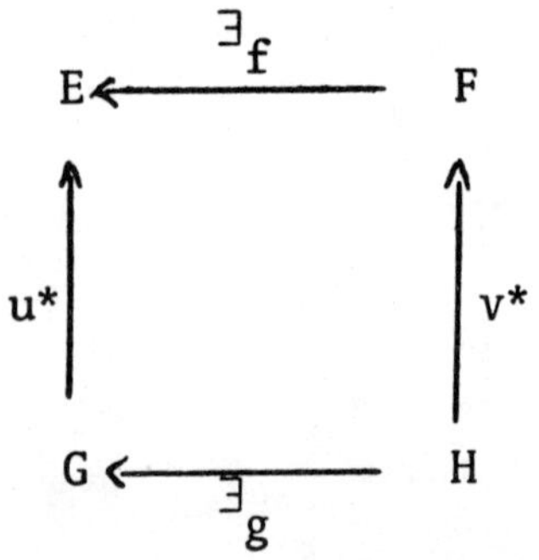

is commutative and that u^* *has a left adjoint* $\exists_u$. *If* f^* *and* u^* *preserve* $\Rightarrow$ *and* $\tau^* = f^* \circ u^*$, *then* $\forall x \in E \quad \forall y \in H \quad \exists_u(x) \wedge \exists_g(y) = \exists_\tau(f^*(x) \wedge v^*(y))$

Proof:

$$\exists_\tau(f^*(x) \wedge v^*(y)) = \exists_u \exists_f(f^*(x) \wedge v^*(y)) = \exists_u(x \wedge \exists_f(v^*(y))),$$

since f^* preserves $\Rightarrow$.

Hence

$$\exists_\tau(f^*(x) \wedge v^*(y)) = \exists_u(x \wedge u^*\exists_g(y)) = \exists_u(x) \wedge \exists_g(y),$$

since u^* preserves $\Rightarrow$.

Definition 2.7: A *regular category with stable* $\vee$ is a regular category $\mathcal{A}$ such that $\mathcal{P}(A)$ is a lattice with smallest element 0_A for every $A \in |\mathcal{A}|$ and such that $f^*: \mathcal{P}(B) \to \mathcal{P}(A)$ preserves v and the smallest element 0_A.

Remark 2.8: This implies that $\mathcal{P}(A)$ is a distributive lattice with smallest and greatest element. Furthermore, $\exists_f: \mathcal{P}(A) \to \mathcal{P}(B)$ preserves v and smallest element since f^* is a right adjoint.

Theorem 2.9: *The following pull-back is a push-out if*

$$A = B_1 \vee B_2$$

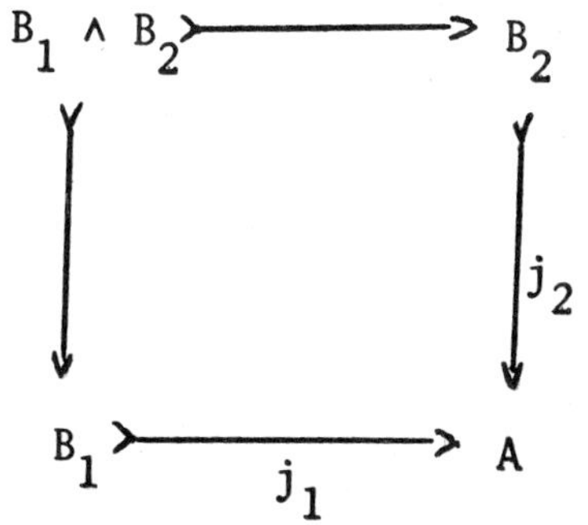

Proof: Suppose that the following commutative diagram is given:

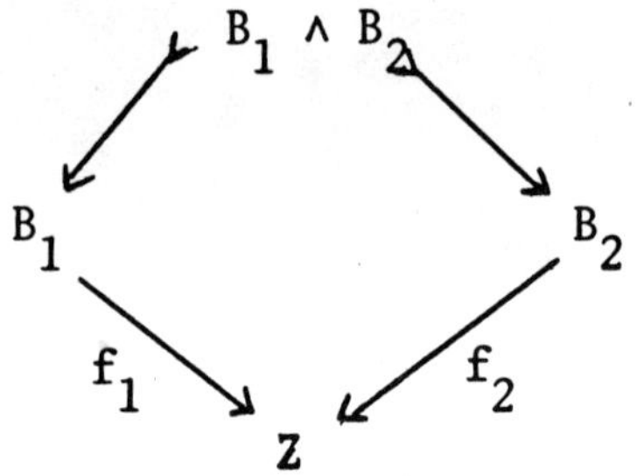

Consider the following diagram:

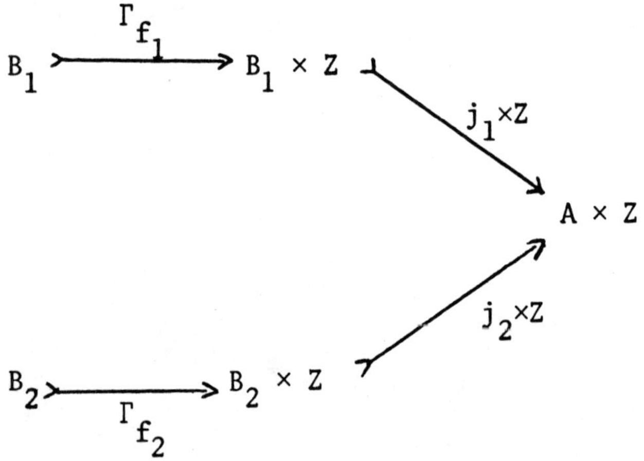

Define $$\Gamma^1_{f1} = Im((j_1 \times Z) \circ \Gamma_{f_1})$$

$$\Gamma^1_{f2} = Im((j_2 \times Z) \circ \Gamma_{f_2})$$

and let α be the composition

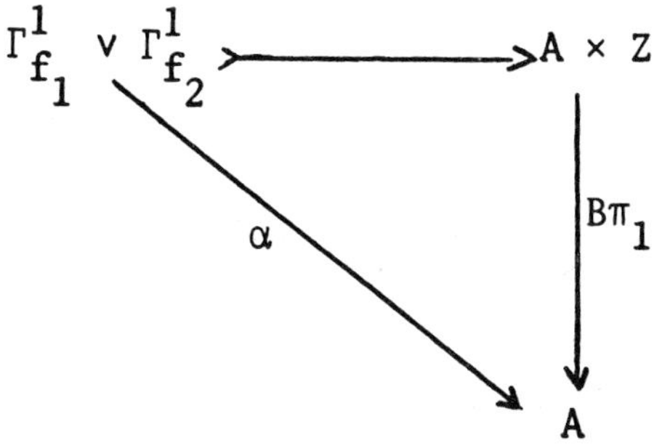

We shall prove that α is an isomorphism. This will allow us to define $g: A \to Z$ such that

$$(B_i \rightarrowtail A \xrightarrow{g} Z) = (B_i \xrightarrow{f_i} Z) \quad \text{by letting}$$

$$g = A \xrightarrow{\alpha^{-1}} \Gamma^1_{f1} \vee \Gamma^1_{f2} \rightarrowtail A \times Z \xrightarrow{\pi_2} Z.$$

Uniqueness of g is obvious.

We shall prove that

$$\alpha^*(B_1) = \Gamma^1_{f1} \quad \text{and} \quad \alpha^*(B_2) = \Gamma^1_{f2}.$$

Indeed

$$\begin{aligned} \alpha^*(B_1) &= p^*(B_1) \wedge (\Gamma^1_{f1} \vee \Gamma^1_{f2}) \\ &= (B_1 \times Z) \wedge (\Gamma^1_{f1} \vee \Gamma^1_{f2}) \\ &= ((B_1 \times Z) \wedge \Gamma^1_{f1}) \vee ((B_1 \times Z) \wedge \Gamma^1_{f2}). \end{aligned}$$

As we have $B_1 \times Z \supset \Gamma^1_{f1}$, it is enough to prove that

$$(B_1 \times Z) \wedge \Gamma^1_{f2} \subset \Gamma^1_{f1}.$$

With this aim in mind, consider the following pull-back diagram which gives us the sought intersection

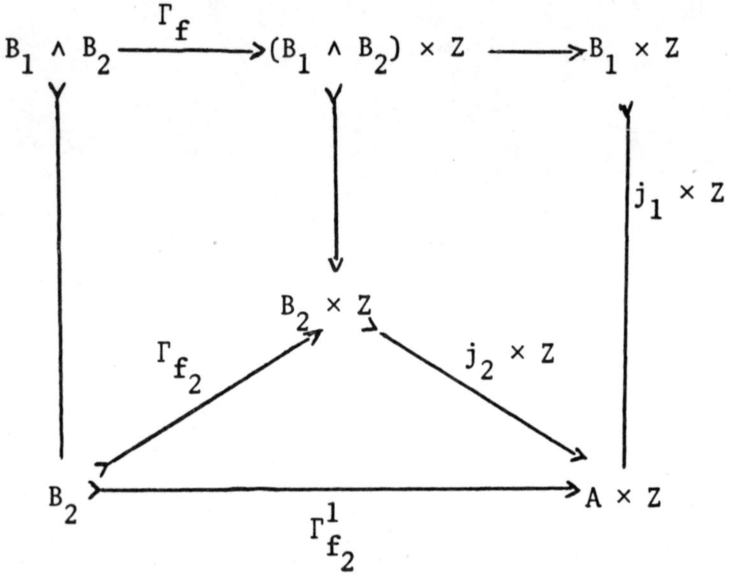

where Γ_f is the graph of the morphism $f\colon B_1 \wedge B_2 \to Z$ obtained by composition

$$(B_1 \wedge B_2 \to B_1 \overset{f_1}{\to} Z) = (B_1 \wedge B_2 \rightarrowtail B_2 \overset{f_2}{\to} Z).$$

The following commutative diagram

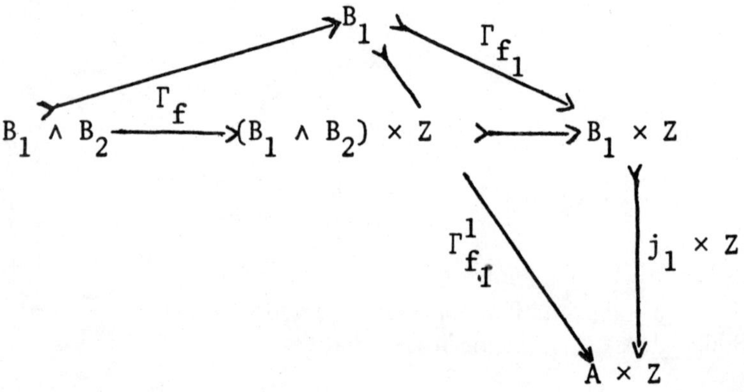

finishes the proof that $\alpha^* B_1 = \Gamma^1_{f_1}$. Moreover, the preceding argument shows that in the pull-back

$$\begin{array}{ccc} \Gamma^1_{f_1} & \rightarrowtail & \Gamma^1_{f_1} \vee \Gamma^1_{f_2} \\ \downarrow & & \downarrow \alpha \\ B_1 & \rightarrowtail & X \end{array}$$

the left arrow $\Gamma^1_{f_1} \to B_1$ is an isomorphism. The proof will be complete once that we have established the following:

LEMMA 2.10: *Suppose that $\alpha: Y \to X$ is a morphism which has the following property: there are two subobjects $B_1 \rightarrowtail X$, $B_2 \rightarrowtail X$ whose supremum $B_1 \vee B_2 = X$ and such that the left arrow of the two following pull-backs is an isomorphism:*

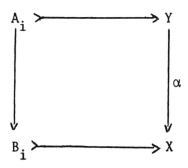

$$\begin{array}{ccc} A_i & \rightarrowtail & Y \\ \downarrow & & \downarrow \alpha \\ B_i & \rightarrowtail & X \end{array}$$

Then α is an isomorphism.

Proof: α is a regular epimorphism since $Im(\alpha) \supseteq B_1 \vee B_2 = X$.

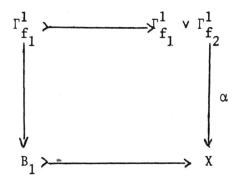

Suppose that $W \overset{u}{\underset{v}{\rightrightarrows}} Y$ are such that $\alpha u = \alpha v = h$. We shall prove that $Ker(u, v) = W$ by showing that $Ker(u, v) \supseteq h^*(B_i)$ since $h^*(B_1) \vee h^*(B_2) = h^*(B_1 \vee B_2) = h^*(X) = W$.

But $Ker(u, v) \supseteq h^*(B_i)$ is equivalent to

$$u \mid h^*(B_i) = v \mid h^*(B_i).$$

The diagram

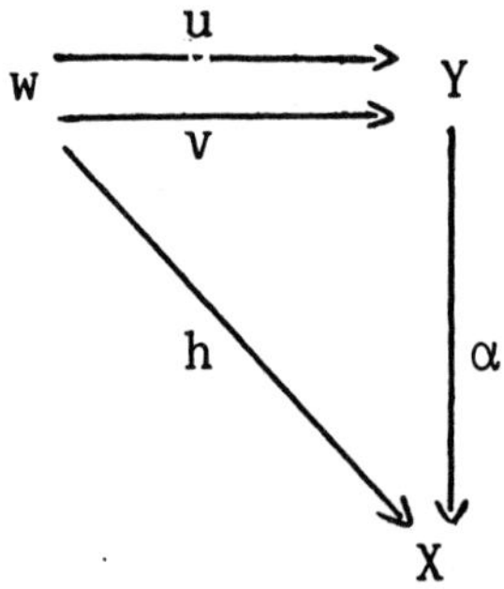

is a diagram in the category $\mathfrak{A}/X$. Pulling it back along $B_i \rightarrowtail X$ we obtain

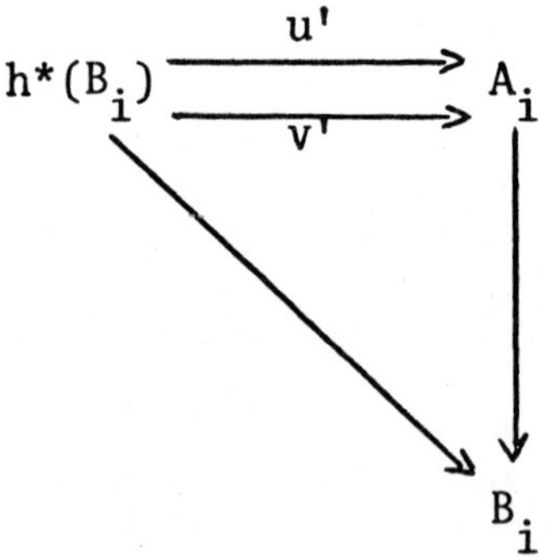

where the right vertical arrow is an isomorphism. We conclude that $u' = v'$, i.e., $u \mid h^*(B_i) = v \mid h^*(B_i)$.

DEFINITION 2.11: A *regular category with* $\Rightarrow$ is a regular category $\mathfrak{A}$ such that for every $A \in |\mathfrak{A}|$ $\mathcal{P}(A)$ is an implicative semilattice.

PROPOSITION 2.12: *In a regular category with* $\Rightarrow$, $f^*\colon \mathcal{P}(B) \to \mathcal{P}(A)$ *preserves* $\Rightarrow$, *for every* $A \xrightarrow{f} B$.

DEFINITION 2.13: A *regular category with* $\forall$ is a regular category $\mathcal{A}$ such that for every $A \xrightarrow{f} B$, $f^*\colon \mathcal{P}(B) \to \mathcal{P}(A)$ has a right adjoint $\forall_f$.

PROPOSITION 2.14: *Every regular category with* $\forall$ *is a regular category with* $\Rightarrow$.

Proof: If $S \overset{j}{\rightarrowtail} A$ and $T \rightarrowtail A$ are subobjects of A, define $(S \Rightarrow T) = \forall_j(S \wedge T)$.

PROPOSITION 2.15: *For every pull-back*

$$\begin{array}{ccc} A' & \xrightarrow{\;f\;} & A \\ {\scriptstyle u'}\downarrow & & \downarrow{\scriptstyle u} \\ B' & \xrightarrow[\;g\;]{} & B \end{array}$$

in a regular category with $\forall$, *the following diagram commutes:*

$$\begin{array}{ccc} P(A') & \xleftarrow{\;f^*\;} & P(A) \\ {\scriptstyle \forall_{u'}}\downarrow & & \downarrow{\scriptstyle \forall_u} \\ P(B') & \xleftarrow[\;g^*\;]{} & P(B) \end{array}$$

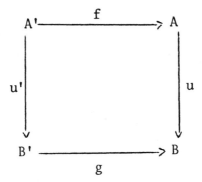

Proof: See Proposition 1.13 and Corollary 2.3.

PROPOSITION 2.16: *A regular category* $\mathcal{A}$ *with* $\forall$ *is a regular category with stable* $\vee$ *iff for every* $A \in |\mathcal{A}|$ $\mathcal{P}(A)$ *is a lattice with smallest element.*

Proof: If $f^*: \mathcal{P}(B) \to \mathcal{P}(A)$ has a right adjoint, then it preserves supremum and smallest element whenever they exist.

DEFINITION 2.17: A *Heyting category* is a regular category with $\forall$ and (stable) $\vee$.

DEFINITION 2.18: A *Boolean category* is a Heyting category for which $\mathcal{P}(A)$ is a Boolean algebra for every object A.

PROPOSITION 2.19: *A regular category with stable* $\vee$ *is Boolean iff the distributive lattice* $\mathcal{P}(A)$ *is complemented.*

Each of the notions of enriched regular category studied so far gives rise to a category. Thus the category of regular categories has *regular functors* as morphisms, i.e., functors which preserve finite $\lim_{\leftarrow}$ and images. The category of Heyting categories has *Heyting* functors as morphisms, i.e., regular functors preserving $\vee$, $\Rightarrow$, $\forall_f$. Similarly, for the other notions. Boolean functors from a Boolean category into the category S of sets will be called *Boolean models.*

SECTION 2

SITES AND SHEAVES

In this chapter, we recall some results of the theory of topoi and sheaves. Proofs and further developments can be found in [V] and [KW]. We assume that all categories involved have fibered products, which allows us to define topologies in terms of covering families. We shall omit references to "universes" to simplify the exposition.

DEFINITION 2.1: A *topology* on $\mathcal{A}$ is an operator Cov such that

$\forall A \in |\mathcal{A}|$, $\mathrm{Cov}(A)$ is a class of families of morphisms of a co-domain A satisfying the following axioms:

(T1) If $\{A_i \to A\}_{i \in I} \in \mathrm{Cov}(A)$, then

$$\forall B \to A \in \mathcal{A}, \qquad \{A_i \times_A B \to B\}_{i \in I} \in \mathrm{Cov}(B)$$

(stability under change of basis).

(T2) If $\{A_i \to A\}_{i \in I} \in \mathrm{Cov}(A)$ and $\{A_{ij} \to A_i\}_{j \in J_i} \in \mathrm{Cov}(A_i)$, then

$$\{A_{ij} \to A\}_{\substack{i \in I \\ j \in J_i}} \in \mathrm{Cov}(A)$$

(stability under composition).

(T3) $\{A \xrightarrow{1_A} A\} \in \mathrm{Cov}(A)$.
The elements of $\mathrm{Cov}(A)$ are called the *coverings* of A.

We can compare two topologies Cov, Cov_1 on a given category $\mathcal{A}$; namely

DEFINITION 2.2: Cov_1 *is finer than* Cov if every family $\{A_i \to A\}_{i \in I} \in \mathrm{Cov}(A)$ belongs to $\mathrm{Cov}_1(A)$, for every $A \in |\mathcal{A}|$. Under the same circumstances, we say that Cov *is coarser than* Cov_1.

We remark that if $(\mathrm{Cov}_i)_{i \in I}$ is a family of topologies on $\mathcal{A}$, the operator Cov defined by $\mathrm{Cov}(A) = \bigcap_{i \in I} \mathrm{Cov}_i(A)$ is again a topology which is the infimum of $(\mathrm{Cov}_i)_{i \in I}$. We shall write $\mathrm{Cov} = \bigwedge_{i \in I} \mathrm{Cov}_i$.

We now give some *examples* of topologies:

(1) Let $\mathrm{Cov}(A)$ be the class of all families $\{A_i \to A\}_{i \in I}$. This is called the *discrete topology* on $\mathcal{A}$.

(2) Let $\mathrm{Cov}(A)$ consist of $\{A \xrightarrow{1_A} A\}$ only. This is called the *chaotic topology* on $\mathcal{A}$.

These two examples are extreme: the first is the finest possible topology, whereas the second is the coarsest possible topology.

The following example will be of interest later.

(3) Let $\mathfrak{A}$ be a Heyting category, and let $\mathrm{Cov}(A)$ be defined as the set of families $\{A_i \xrightarrow{f_i} A\}_{i \in I}$ such that for every $C \to A \in \mathfrak{A}$ where $C \neq 0_C$ there is some $D \neq 0_D$ and morphisms $D \to C \in \mathfrak{A}$ and $D \to A_i \in \mathfrak{A}$ for some $i \in I$ making the diagram

$$\begin{array}{ccc} A_i & \xrightarrow{f_i} & A \\ \uparrow & & \uparrow \\ D & \longrightarrow & C \end{array}$$

commutative. This topology is called the *double negation topology*.

Definition 2.3: A *site* is a category together with a topology. A morphism of sites is a continuous functor between the underlying categories, i.e., one preserving fiber products and covering families.

Whenever we have a functor u from a category $\mathfrak{A}$ into a site $\mathfrak{B}$ which preserves fiber products, we can define:

$$\{A_i \to A\}_{i \in I} \in \mathrm{Cov}(A) \quad \text{iff} \quad \{uA_i \to uA\}_{i \in I} \in \mathrm{Cov}(uA).$$

It is easy to check that Cov is a topology on $\mathfrak{A}$.

Definition 2.4: The topology defined above is called the *induced topology*.

We now arrive at the main concept of this chapter.

Definition 2.5: Let $\mathfrak{A}$ be a site. A *pre-sheaf* over $\mathfrak{A}$ is a contravariant functor from $\mathfrak{A}$ into the category of sets $\mathcal{S}$. A pre-sheaf F is a *sheaf* if for every $A \in |\mathfrak{A}|$ and every covering $\{A_i \to A\}_{i \in I} \in \mathrm{Cov}\, A$ the obvious $\mathcal{S}$-diagram

$$F(A) \to \prod_{i \in I} F(A_i) \rightrightarrows \prod_{i,j \in I} F(A_i \underset{A}{\times} A_j)$$

is exact, i.e., $F(A)$ is the equalizer of the pair of maps.

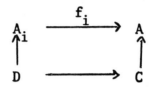

We shall denote by $\hat{\mathfrak{A}}$ the category of pre-sheaves (with natural transformations as morphisms) and by $\tilde{\mathcal{A}}$ the full subcategory of sheaves. Notice that this definition makes sense whenever $\mathcal{A}$ is a small category.

Intuitively, we can say that sheaves are set-like objects for which "partial functions into" them which are compatible and whose domain cover an object give rise to a unique "total function on the domain." Furthermore, they are closed under several "set-theoretic" operations, as the following shows:

PROPOSITION 2.6: *Let $\mathcal{B}$ be the category of sheaves over a (small) site. Then the following conditions are satisfied:*

(1) *$\mathcal{B}$ is a Heyting category.*
(2) *$\mathcal{B}$ is co-complete and* $\lim_{\rightarrow}$ *are universal in the following sense:*

$$\text{if } A = \lim_{\substack{\rightarrow \\ i\in I}} A_i, \text{ then } \forall B \rightarrow A \in \mathcal{B} \quad B = \lim_{\substack{\rightarrow \\ i\in I}} A_i \times_A B.$$

(3) *For every $A \in |\mathcal{B}|$ the functor $-\times A\colon \mathcal{B} \rightarrow \mathcal{B}$ has a right adjoint $(-)^A$.*
(4) *The canonical injections $A_i \rightarrow \coprod_{i\in I} A_i$ are monomorphisms*

and $A_i \times_A A_j = 0$, where $i \neq j$ and $S = \coprod_{i\in I} A_i$.

As we saw in the introduction, the "logic of the circle" is not Boolean. Hence, we cannot demand $\mathcal{B}$ Boolean. However, the following holds:

PROPOSITION 2.7: *Let $\mathcal{A}$ be a Heyting category with the double negation topology. Then $\tilde{\mathcal{A}}$ is Boolean.*

DEFINITION 2.8: A *topos* is a category equivalent to the category of sheaves over some (small) site.

PROPOSITION 2.9: *On any topos, let* Cov *be defined by*

$$\{A_i \xrightarrow{f_i} A\}_{i\in I} \in \mathrm{Cov}(A) \quad \text{iff} \quad V_{i\in I}\, Im(f_i) = A.$$

Then Cov *is the finest topology for which all representable functors are sheaves.*

Definition 2.10: Let $\mathfrak{A}$ be any category (with fiber products). A topology is *canonical* if it is the finest topology for which all representable functors are sheaves.

The following proposition tells us that there is always a "unique" canonical topology:

Proposition 2.11: *Any category (with fiber products) has a (unique) canonical topology.*

The following result (Corollary of a theorem of Giraud) allows us to recover a topos from certain small subcategories with the induced topology:

Proposition 2.12: *Assume that $\mathfrak{B}_0 \to \mathfrak{B}$ is a full small subcategory of a topos $\mathfrak{B}$ which has fiber products and is closed under subobjects. Then the induced topology on $\mathfrak{B}_0$ by the canonical topology on $\mathfrak{B}$ is the canonical topology on $\mathfrak{B}_0$. Furthermore, $\tilde{\mathfrak{B}}_0 \simeq \mathfrak{B}$, provided that $\mathfrak{B}_0$ contains a set of generators of $\mathfrak{B}$.*

Let $\mathfrak{A}$ be a site. For every $A \in |\mathfrak{A}|$ and every $\mathfrak{U} = \{A_i \to A\} \in \operatorname{Cov}(A)$, we define $R_{\mathfrak{u}} \rightarrowtail h_A$ as follows: $R_{\mathfrak{u}}(B) = \{B \xrightarrow{f} A \in \mathfrak{A} \mid f$ factors through some $A_i \to A \in \mathfrak{U}\}$.

Furthermore, we let $J(A)$ be the class of all functors $R \rightarrowtail h_A$ such that $R_{\mathfrak{u}} \rightarrowtail R$, for some $\mathfrak{U} \in \operatorname{Cov}(A)$.

There is a natural structure of (ordered) category on $J(A)$ (i.e., there is a unique morphism $R \to R'$ iff $R \rightarrowtail R'$). We notice that $J(A)^{opp}$ is a filtered category.

Definition 2.13: If F is a pre-sheaf on a site $\mathfrak{A}$, we let $LF \in \hat{\mathfrak{A}}$ be defined by

$$LF(A) = \varinjlim_{R \in J(A)} \hat{\mathfrak{A}}_{opp}(R, F).$$

We can consider L as a functor from $\hat{\mathfrak{A}}$ into $\hat{\mathfrak{A}}$ in an obvious way.

Furthermore, define $a = L \circ L$. Then $aF \in \tilde{\mathfrak{A}}$, for every $F \in \mathfrak{A}$. In fact, more is true as shown by the fundamental

Theorem 2.14: *Let $\mathfrak{A}$ be a small site, $\tilde{\mathfrak{A}}$ the category of sheaves and $i: \tilde{\mathfrak{A}} \to \hat{\mathfrak{A}}$ the inclusion functor into the category of pre-sheaves. Then $a: \hat{\mathfrak{A}} \to \tilde{\mathfrak{A}}$ is a left adjoint to i which commutes with finite* $\varprojlim$.

Let $\mathfrak{A}$, $\mathfrak{A}'$ be two categories, $F: \mathfrak{A} \to \mathfrak{A}'$ a functor. We denote by $F_\bullet: \hat{\mathfrak{A}}' \to \hat{\mathfrak{A}}$ the functor defined by

$$(F_\bullet G)(A) = G(F(A)), \quad \text{for all} \quad A \in |\mathfrak{A}|.$$

Similarly, we shall denote by $F^\bullet: \mathfrak{A} \to \mathfrak{A}'$ the Kan extension of F which is a left adjoint to $F_\bullet$. We recall that if $\hat{\mathfrak{A}}$ has finite $\lim_\leftarrow$ and F commutes with these finite $\lim_\leftarrow$, then so does $F^\bullet$.

PROPOSITION 2.15: *Let $\mathfrak{A}$, $\mathfrak{A}'$ be two sites, let $F: \mathfrak{A} \to \mathfrak{A}'$ be a continuous functor. Let i, i' be the canonical injections $\tilde{\mathfrak{A}} \to \hat{\mathfrak{A}}$, $\tilde{\mathfrak{A}}' \to \hat{\mathfrak{A}}'$. We define $F_* = F_\bullet \circ i'$ and $F^* = a' \circ F^\bullet \circ i$. Then $F_*(G') \in \tilde{\mathfrak{A}}$, for all $G' \in \tilde{\mathfrak{A}}'$ and $F^*: \tilde{\mathfrak{A}} \to \tilde{\mathfrak{A}}'$ is a left adjoint to F_*. Furthermore, if $\mathfrak{A}$ has finite* $\lim_\leftarrow$ *and F commutes with these, then so does F^*.*

We finish with the definition of the category of topoi:

DEFINITION 2.16: Let $\mathfrak{A}$, $\mathfrak{B}$ be topoi. A *morphism* of topoi is a couple (U, V) of functors $\mathfrak{A} \underset{V}{\overset{U}{\rightleftarrows}} \mathfrak{B}$ such that $U \dashv V$ and U commutes with finite $\lim_\leftarrow$.

COROLLARY 2.17: *Assume that $\mathfrak{A}$, $\mathfrak{A}'$ are small sites having finite* $\lim_\leftarrow$ *and $F: \mathfrak{A} \to \mathfrak{A}'$ a continuous functor preserving finite* $\lim_\leftarrow$. *Then*

$$\tilde{\mathfrak{A}} \underset{F_*}{\overset{F^*}{\rightleftarrows}} \tilde{\mathfrak{A}}'$$

is a morphism of topos.

SECTION 3

FROM SHEAVES TO LOGIC

In this section, we compare the "geometric" and logical aspects of a category of sheaves. More precisely, any topos has the logical structure of a Heyting category (II.2.6). This means that the logical operations $\wedge$, $\vee$, $\Rightarrow$, $\exists$, $\forall$ are definable and satisfy the laws of intuitionistic logic. Furthermore, it is clear that any choice of logical operations (definable in the topos) gives rise to a corresponding logical structure. Choosing $\wedge, \exists, \vee$ as our logical operations, for instance, the corresponding logical structure is the topos considered as a regular category with stable $\vee$. In this fashion, we obtain a forgetful functor from the category of topoi to the category of regular categories with stable $\vee$.

We shall study the problem of the existence of left adjoints to this forgetful functor, for different choices of logical operations. This is one aspect of the problem of "solving the contradiction" between geometry and logic, emphasized by Lawvere [L3].

To formulate our results succinctly, we shall ignore set-theoretical difficulties.

In §1, we construct a left adjoint for the choice $\wedge, \exists, \vee$.

If we enrich our choice to $\wedge, \exists, \vee, \Rightarrow$ and we consider the category of topoi with morphisms (U, V), and such that U preserves $\Rightarrow$, then the forgetful functor does not possess any left adjoint. This is shown in §2.

In §3, we consider a "compromise solution." We restrict our attention to the category of Boolean topos and we choose all first-order operations $\wedge, \exists, \vee, \Rightarrow, \forall$. Then we show that the forgetful functor into the category of Boolean categories with *continuous* Boolean functors (i.e., preserving arbitrary sups) has a left adjoint. As D. Higgs has shown, the category of Boolean valued sets appears as a particular instance of this construction. Notice, however, that continuity is a "geometric" notion which has been transferred from the geometric side.

1. FINITE COVER TOPOLOGY

Let $\mathfrak{a}$ be a regular category with stable $\vee$. For every $A \in |\mathfrak{a}|$, we let $\mathrm{Cov}(A)$ be the class of all finite families $\{A_i \xrightarrow{f_i} A\}_{i\in I}$ such that $\vee_{i\in I} Im(F_i) = A$.

PROPOSITION 1.1: Cov *is a topology on* $\mathfrak{a}$.

Definition 1.2: This topology is called the *finite cover topology.*

Proposition 1.3: *The finite cover topology is coarser than the canonical topology, i.e., the representable functors are sheaves.*

Proof: Let $\{A_i \xrightarrow{f_i} A\}_{i \in I} \in \mathrm{Cov}(A)$ and let $C \in |\mathcal{A}|$. Furthermore, assume that $(\xi_i)_{i \in I}$ belongs to the equalizer of

$$\prod_{i \in I} h_C(A_i) \rightrightarrows \prod_{i,j \in I} h_C(A_i \times_A A_j).$$

We have to show the existence of $\xi \in h_C(A)$ such that $\xi \circ f_i = \xi_i$ for all $i \in I$.

By induction, we may assume that $I = \{0, 1\}$.

Let $A_i \twoheadrightarrow A_i' \rightarrowtail A$ be the unique factorization of f_i in a regular epimorphism followed by a monomorphism. By theorem I.1.21 there is a unique morphism $A_i' \xrightarrow{\xi_i'} C$ such that

$$\xi_i = A_i \twoheadrightarrow A_i' \xrightarrow{\xi_i'} C \quad \text{for} \quad i = 0, 1.$$

From Corollary, I.2.6, the diagram

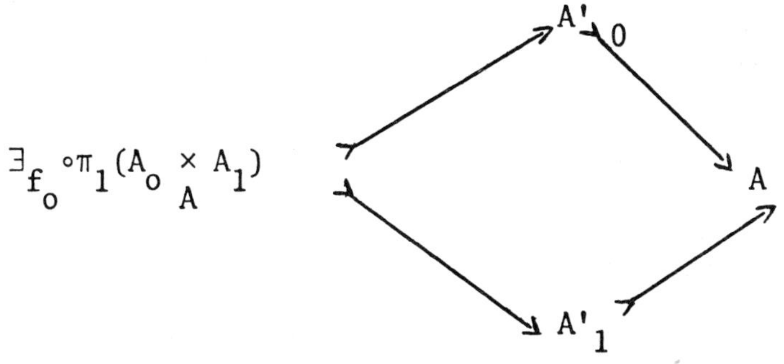

is a pull-back. Furthermore, by Theorem I.2.9 our previous diagram is also a push-out and this shows the existence of a unique $A \xrightarrow{\xi} C$ with the desired properties.

From now on, we let $\tilde{\mathcal{A}}$ be the category of sheaves for the finite cover topology.

Proposition 1.4: *Let $\mathcal{A}$ be a regular category with stable $\vee$. Then the Yoneda functor $h\colon \mathcal{A} \rightarrowtail \tilde{\mathcal{A}}$ preserves finite* $\lim_{\leftarrow}$, *regular epimorphisms and $\vee$. Furthermore, h preserves any of the operations $\Rightarrow$, $\forall$ which exist in $\mathcal{A}$.*

Proof: Since the Yoneda functor $h\colon \mathcal{A} \rightarrowtail \tilde{\mathcal{A}}$ preserves arbitrary projective limits and since projective limits in $\tilde{\mathcal{A}}$ are computed pointwise, then $h\colon \mathcal{A} \rightarrowtail \tilde{\mathcal{A}}$ preserves all projective limits which exist in $\mathcal{A}$.

Let $A \overset{f}{\twoheadrightarrow} B$ be a regular epimorphism and assume that f equalizes the pair $B \overset{\xi}{\underset{\eta}{\rightrightarrows}} F$, where $F \in \tilde{\mathcal{A}}$. We thus obtain a commutative diagram

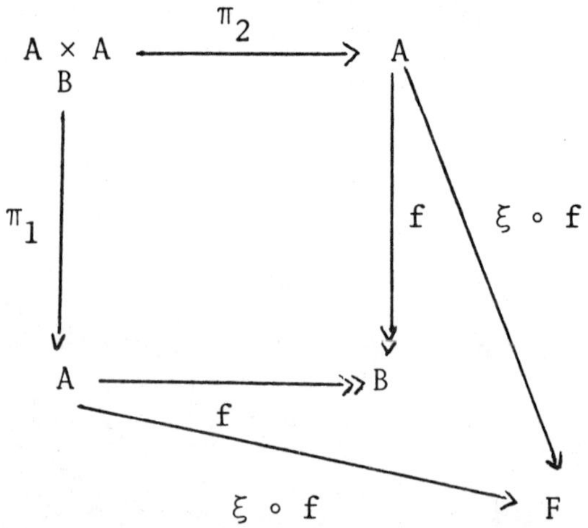

Since $\{A \overset{f}{\twoheadrightarrow} B\} \in \mathrm{Cov}(B)$ for the finite cover topology and F is a sheaf for this topology, there is a unique $\theta\colon B \to F$ such that $\theta \circ f = \xi \circ f$. This implies that $\xi = \eta = \theta$ and so $A \overset{f}{\twoheadrightarrow} B$ is epic in $\tilde{\mathcal{A}}$.

We claim that $h_{A \vee B} = h_A \coprod_{h_{A \wedge B}} h_B$. In fact, consider the commutative diagram

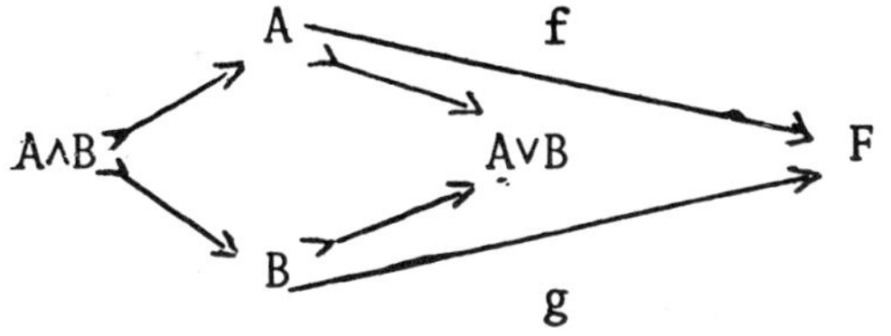

But $\{f, g\} \in \mathrm{Cov}(A \vee B)$ and F is a sheaf, i.e., there is a unique $h: A \vee B \to F$ making the previous diagram commutative. By Theorem I.2.9 $h_{A \vee B} = h_A \vee h_B$, since $h_{A \wedge B} = h_A \wedge h_B$.

To finish the proof, we need the following

LEMMA 1.5: *Let $A \xrightarrow{f} B \in \mathfrak{A}$ be such that the pulling-back functor*

$$\mathfrak{A}/B \xrightarrow{*f} \mathfrak{A}/A$$

has a right adjoint Π_f. Then for all $\delta \in |\mathfrak{A}/A|$,

$$h_{\Pi_f(\delta)} = \Pi_f(h_\delta).$$

Proof: Let $F \in |\tilde{\mathfrak{A}}|$ and consider the following diagram

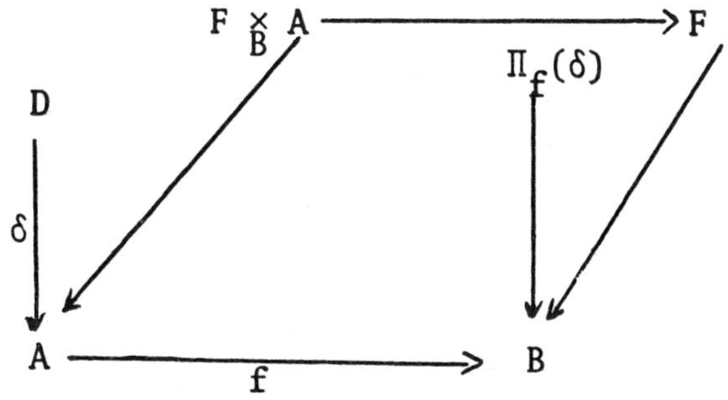

Then

$$\begin{aligned}\tilde{\mathcal{A}}/B(F, \Pi_f(\delta)) &= \tilde{\mathcal{A}}/B(\lim_{\to C \to F} C, \Pi_f(\delta)) \\ &= \lim_{\leftarrow C \to F} \mathcal{A}/B(C, \Pi_f(\delta)) \\ &\simeq \lim_{\leftarrow C \to F} \mathcal{A}/A(A \underset{B}{\times} C, \delta) \\ &= \tilde{\mathcal{A}}/A(\lim_{\to C \to F} A \underset{B}{\times} C, \delta).\end{aligned}$$

Since $\lim_{\to}$ are universal in a topos (II.2.6) $\lim_{\to C \to F} A \underset{B}{\times} C = A \underset{B}{\times} F$ this concludes the proof of the lemma.

We observe, furthermore, that if $D \overset{\delta}{\rightarrowtail} A$, then $\Pi_f(\delta) = \forall_f(D)$ since $\Pi_f(D) \to B$ is a monomorphism in this case and the adjointness condition reduces to

$$\frac{C \longrightarrow \Pi_f(D)}{f^{-1}(C) \rightarrowtail D}$$

This concludes the proof of the proposition.

Remark 1.6:

(i) The finite cover topology is the coarsest topology on a regular category with stable $\vee$ for which the Yoneda embedding preserves finite $\lim_{\leftarrow}$, images and $\vee$.

We now define a functor $\sim$ from the category of regular categories with stable $\vee$ into the category of topoi as follows:

(i) $\tilde{\mathcal{A}}$ is the category of sheaves for the finite cover topology.

(ii) Let $\mathcal{A} \overset{u}{\to} \mathcal{B}$ be a regular functor. Since u is continuous for the finite cover topology, II.2.17 gives a morphism of

topos $\tilde{\mathcal{A}} \underset{u_s}{\overset{u^s}{\rightleftarrows}} \tilde{\mathcal{B}}$ such that $u^s \dashv u_s$, which we define to be $\tilde{u}$.

THEOREM 1.7: *The functor* $\sim$ *is a left adjoint to the forgetful functor. More precisely, let* $\mathcal{A}$ *be a regular category with stable* $\vee$ *and let* $h: \mathcal{A} \rightarrowtail \tilde{\mathcal{A}}$ *be the Yoneda embedding into the category of sheaves for the finite cover topology. Then for every functor* $\mathcal{A} \xrightarrow{u} \mathcal{B}$ *into a topos such that u preserves finiie* $\varprojlim$, *images and* $\vee$, *there is a unique (up to isomorphism) morphism of topos*

$$\tilde{\mathcal{A}} \underset{\underset{\sim}{u}}{\overset{\tilde{u}}{\rightleftarrows}} \mathcal{B}$$

such that

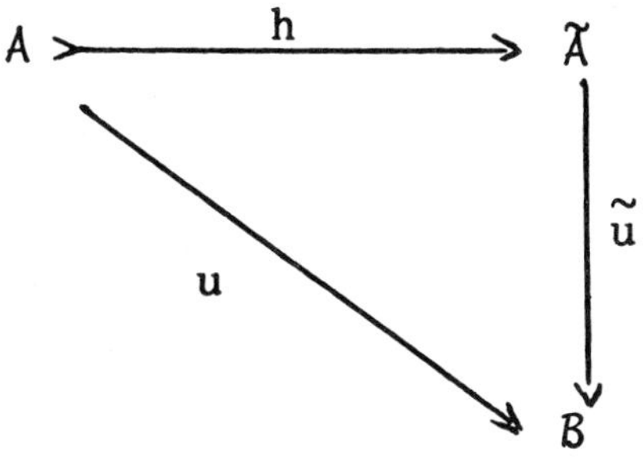

is commutative.

Proof: Let $\mathcal{B}_0 \rightarrowtail \mathcal{B}$ be a small full subcategory of $\mathcal{B}$ such that

(i) $|\mathcal{B}_0| \supset |u(\mathcal{A})|$,
(ii) $|\mathcal{B}_0|$ contains a set of generators of $\mathcal{B}$,
(iii) $\mathcal{B}_0$ has finite $\varprojlim$, images, $\vee$,
(iv) $\mathcal{B}_0$ is closed under sub-objects.

It is clear then that $\mathcal{B}_0$ is a regular category with stable $\vee$.

Furthermore we have the following commutative diagram

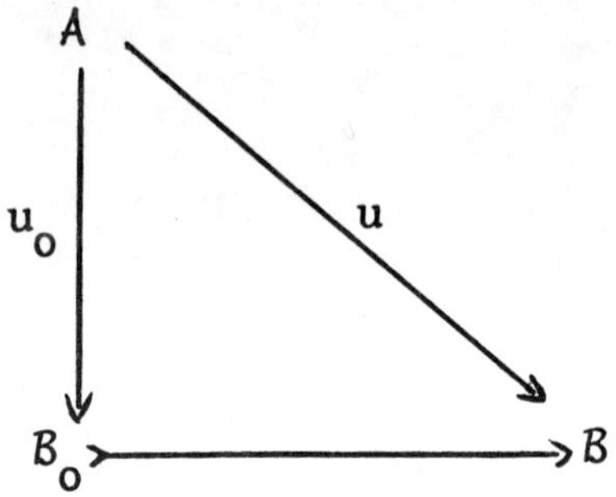

where u_0 preserves finite $\lim_{\leftarrow}$, images, $\vee$. By II.2.1 the induced topology on $\mathcal{B}_0$ by the canonical topology is the canonical topology and this shows that u_0 is continuous. We thus obtain a morphism of topoi

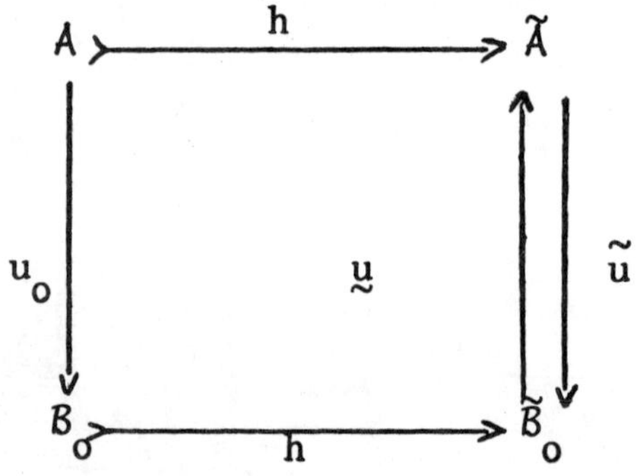

By II.2.12, $\tilde{\mathcal{B}}_0 \cong \mathcal{B}$ finishing the proof.

Let us remark that $\tilde{u}$ preserves not only finite $\lim_{\leftarrow}$, but also images and $\vee$, since images are co-equalizers and $\vee$ are fibered sums, i.e., particular case of $\lim_{\rightarrow}$.

Remark 1.8:

(i) The functor $\sim$ is not conservative, in general. However, we have the following result:

PROPOSITION: *Let* $\mathcal{A} \xrightarrow{u} \mathcal{B}$ *be a regular* $\vee$*-preserving functor between small regular categories with stable* $\vee$. *Assume that* $\mathcal{A}$ *is exact (i.e., the equivalence relations are kernel pairs) and has finite sums* $A = \coprod_{i \in F} A_i$ *such that*

(i) *the canonical injections* $A_i \rightarrow A$ *are monomorphisms;*
(ii) $A_i \underset{A}{\times} A_j = 0_A$ *for* $i \neq j$;

Then the following are equivalent:

(1) *u is an equivalence (of categories),*

(2) $\tilde{\mathcal{A}} \underset{u_*}{\overset{\tilde{u}}{\rightleftarrows}} \tilde{\mathcal{B}}$ *is an equivalence.*

(ii) The category $\tilde{\mathcal{A}}$ is not Boolean in general, even when $\mathcal{A}$ is Boolean, as the following shows:

PROPOSITION: *Let* $\mathcal{A}$ *be a small Boolean category. Then the following are equivalent:*

(1) $\tilde{\mathcal{A}}$ *is Boolean,*
(2) $P(A)$ *is finite, for all* $A \in |\mathcal{A}|$,
(3) $\mathcal{A} \rightarrowtail \tilde{\mathcal{A}}$ *preserves arbitrary sups.*

(iii) As Lawvere has pointed out to me, if $\mathcal{A} \xrightarrow{u} \mathcal{B}$ is a regular functor preserving $\vee$, then $u_*: \tilde{\mathcal{B}} \rightarrow \tilde{\mathcal{A}}$ preserves filtered $\lim_{\rightarrow}$. However, it seems unlikely that every morphism of topoi (U, V) such that V preserves filtered $\lim_{\rightarrow}$ can be obtained from a $\vee$-preserving regular functor.

2. NON-EXISTENCE OF LEFT ADJOINTS IN THE GENERAL CASE

We first use a theorem of Gaifman-Hales to show that the following problem P_A does not always have a solution: given a Boolean algebra A, find a $\vee$-complete Heyting algebra $U(A)$ and a

Heyting morphism $A \xrightarrow{u} U(A)$ such that for every Boolean homomorphism $h: A \to B$ into a complete Boolean algebra B there is some sup-preserving Heyting morphism $h': U(A) \to B$ such that the following diagram is commutative:

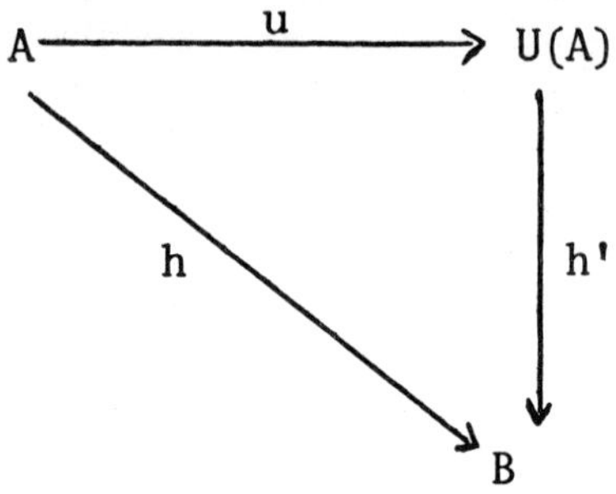

Notice that we do not even require uniqueness for h'.

Let $\mathcal{R}$: Heyting $\to$ Boole be the "retract" functor from the category of Heyting algebras into the category of Boolean algebras defined as follows: if

$$H = (H, \cup, \cap, \Rightarrow, 0.1)$$

is a Heyting algebra, $a \in \mathcal{R}(H)$ iff $a = --a$. The operations on $\mathcal{R}(H)$ are defined by

$$a \wedge b = a \cap b,$$

$$a \wedge b = --(a \cup b),$$

$$\neg a = -a.$$

It is easy to check (or see [RS]) that

$$r: H \to \mathcal{R}(H) \text{ defined by}$$

$$r(a) = --a$$

is a Heyting morphism which is a left adjoint to the inclusion $\mathcal{R}(H) \rightarrowtail H$, when these ordered sets are considered as categories. In particular, if H is $\vee$-complete, then $\mathcal{R}(H)$ is a complete Boolean algebra and

$$V_{i \in I}\, a_i = --\bigcup_{i \in I} a_i.$$

PROPOSITION 2.1: *P_A does not have a solution if A is the free Boolean algebra on $\aleph_0$ generators.*

Proof: Assume, on the contrary, that P_A has a solution $A \xrightarrow{u} U(A)$. Let k be an infinite cardinal. By the theorem of Gaifman-Hales (see [S]) there is a k-Boolean algebra B_k of cardinality at least k and k-generated by a countable set of generators G. The freedom of A assures us of the existence of a Boolean homomorphism $h_0: A \to B_k$ whose image contains G. Letting $\hat{B}_k$ be the McNeille completion of B_k, we thus obtain a Boolean homomorphism $h: A \to B_k \rightarrowtail \hat{B}_k$ into a complete Boolean algebra. Since P_A has a solution, there is some sup-preserving $h'': U(A) \to \hat{B}_k$ such that the diagram

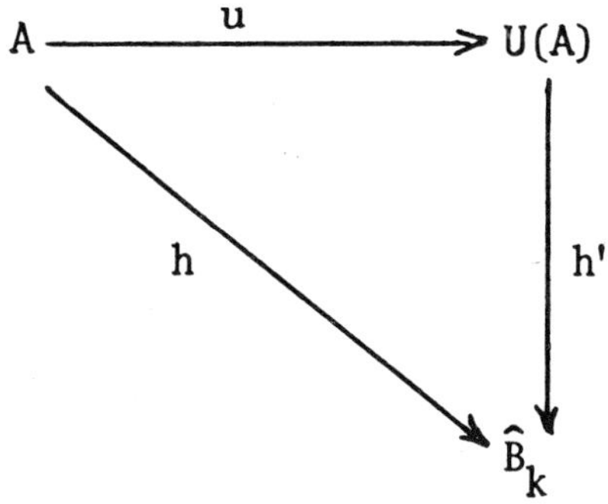

commutes.

The functor $\mathcal{R}$ gives us a commutative diagram

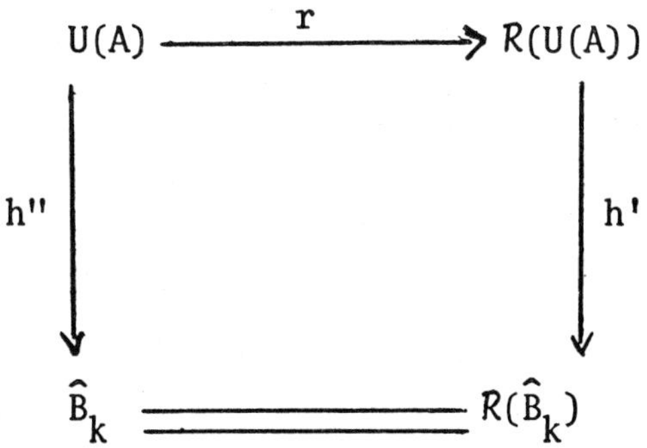

where $\mathcal{R}(U(A))$ is a complete Boolean algebra and $h' = \mathcal{R}(h'')$ is the restriction of h'' to $\mathcal{R}(U(A))$ and hence it is an ∞-homomorphism. Putting our diagrams together, we obtain the commutative triangle

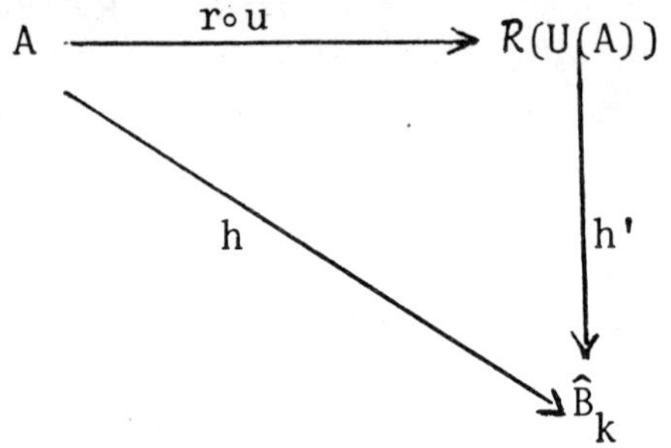

The image of h' is a complete Boolean algebra containing G. In particular, it is a k-Boolean algebra containing G and thus B_k. But B_k is dense in $\hat{B}_k$ and h' is surjective. This shows that the cardinality of $U(A)$ is at least k.

We now "lift" our negative solution to categories. Let $\mathcal{P}_{\mathcal{A}}$ be the following problem: given a Boolean category $\mathcal{A}$, find a topos $\mathcal{U}(\mathcal{A})$ and a functor $\mathcal{A} \xrightarrow{U} \mathcal{U}(\mathcal{A})$ preserving finite $\lim_{\leftarrow}$, images and $\vee$ such that for every Boolean topos $\mathcal{B}$ and every Boolean functor $\mathcal{A} \xrightarrow{F} \mathcal{B}$ there is a functor $\mathcal{U}(\mathcal{A}) \xrightarrow{F'} \mathcal{B}$ which preserves $\lim_{\rightarrow}$, finite $\lim_{\leftarrow}$, $\Rightarrow$ and makes the following diagram commutative

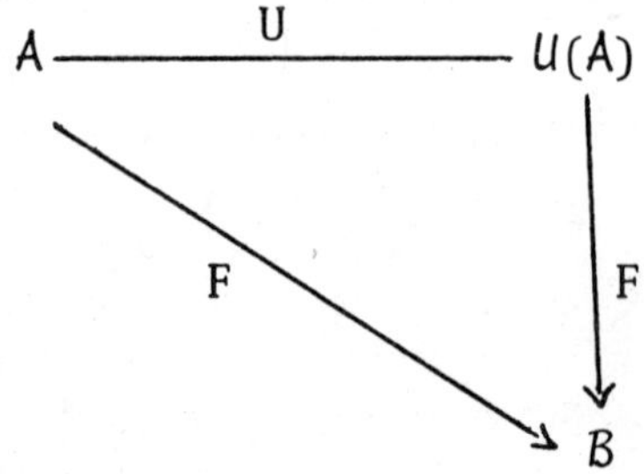

Proposition 2.2: *$\mathcal{P}_A$ has no solution for the free Boolean algebra A generated by $\aleph_0$ generators (considered as a Boolean category).*

Proof: We shall prove that a positive solution of $\mathcal{P}_A$ gives us a positive solution to P_A. Indeed, let $A \xrightarrow{U} \mathfrak{U}(A)$ be a solution of $\mathcal{P}_A$, let $U(A)$ be the Heyting algebra of the sub-objects of 1 in $\mathfrak{U}(A)$ and let $A \xrightarrow{u} U(A)$ be the Heyting morphism such that

$$U = A \xrightarrow{u} U(A) \rightarrowtail \mathfrak{U}(A).$$

We claim that $A \xrightarrow{u} U(A)$ is a solution of P_A. Assume that $A \xrightarrow{f} B$ is a Boolean homomorphism into a complete Boolean algebra B. From the universality of $\mathfrak{U}(A)$ we obtain a functor $F': \mathfrak{U}(A) \to \tilde{B}$ preserving $\Rightarrow$, $\lim_{\to}$, finite $\lim_{\leftarrow}$ and such that

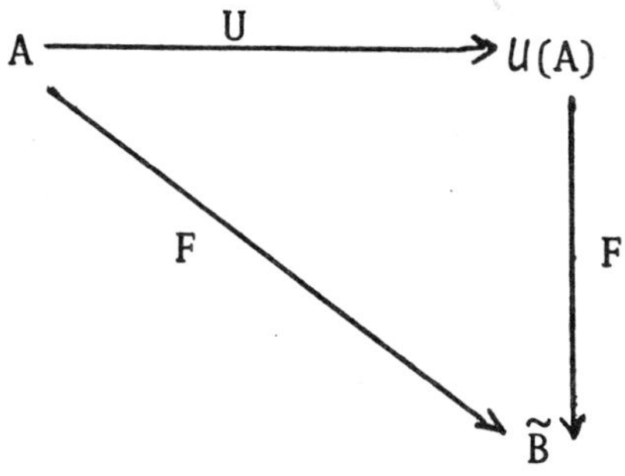

is commutative, where F is the composition of f with the canonical functor $B \rightarrowtail \tilde{B}$ into the category of sheaves for the double negation topology.

By restricting F' to $U(A)$, we obtain a sup-preserving Heyting morphism $U(A) \xrightarrow{f'} B$ such that the diagram

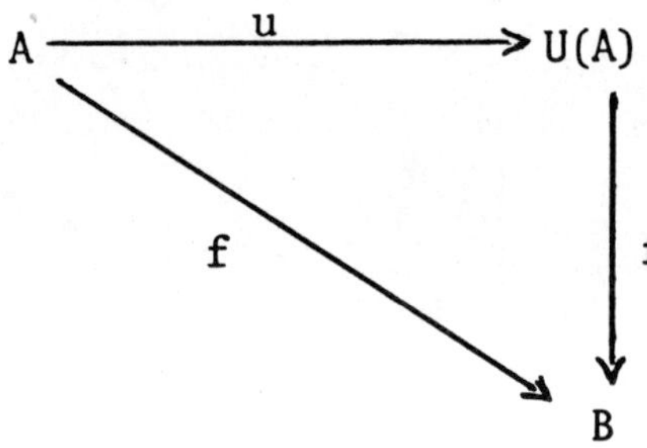

is commutative. In fact, if $A_i \rightarrowtail 1 \forall i \in I$, we have the diagram

$$\coprod_{i \in I} A_i \twoheadrightarrow V_{i \in I} A_i \rightarrowtail 1.$$

Since F' preserves $\lim_{\rightarrow}$, $f' V_{i \in I} A_i = V_{i \in I} f' A_i$. We have shown that $A \xrightarrow{u} U(A)$ is a solution of P_A.

COROLLARY 2.3: *The forgetful functor from the category of topoi with morphisms (U, V) such that U preserves $\Rightarrow$ into the category of regular categories with stable $\vee$ and $\Rightarrow$ does not possess a left adjoint.*

3. DOUBLE NEGATION TOPOLOGY

In this section, we construct a left adjoint $\neg\neg$ to the forgetful functor U from the category of Boolean topos into the category of Boolean categories with continuous (i.e., sup-preserving) Boolean functors.

Let $\mathcal{A}$ be a Boolean category. We make $\mathcal{A}$ into a site by considering the double negation topology (II.2.2). Since $P(A)$ is a Boolean algebra, for each $A \in |\mathcal{A}|$, we can describe the covering families as follows:

$$\{A_i \xrightarrow{f_i} A\}_{i \in I} \in \mathrm{Cov}_{\neg\neg}(A) \quad \text{iff} \quad V_{i \in I} Im(f_i) = A,$$

as easily verified.

We let $\neg\neg\mathcal{A}$ be the Boolean topos of sheaves for the double negation topology.

Furthermore, let $\mathcal{A} \xrightarrow{u} \mathcal{B}$ be a sup-preserving Boolean functor. Then u is continuous for the double negation topology and we thus obtain a morphism of topos

$$\neg\neg\mathcal{A} \underset{u_*}{\overset{u^*}{\rightleftarrows}} \neg\neg\mathcal{B}, \quad \text{such that} \quad u^* \dashv u_*,$$

which we define to be $\neg\neg u$. (See II.2.17.)

In the next proposition, we let $\mathcal{A} \xrightarrow{a} \neg\neg\mathcal{A}$ be the "associated sheaf of" as defined in II.2.13.

PROPOSITION 3.1: *Let* $\mathcal{A}$ *be a small Boolean category. Then* $\mathcal{A} \xrightarrow{a} \neg\neg\mathcal{A}$ *is a Boolean functor which preserves all first-order logical operations* $\wedge$, $\forall$, $\vee$, $\Rightarrow$, $\exists$. *Moreover,* a *preserves* Π_f *whenever it exists.*

Proof: Since the double negation topology is finer than the finite cover topology, a preserves finite $\lim_{\leftarrow}$, images and $\vee$, as can be seen from the proof of III.1.4. Moreover, since $\mathcal{A}$ is Boolean $\Rightarrow$, $\forall$ are definable in the usual way in terms of the other logical operations and so they are preserved by a.

The proof of the last part follows the same lines of the proof of III.1.5. We consider the following diagram

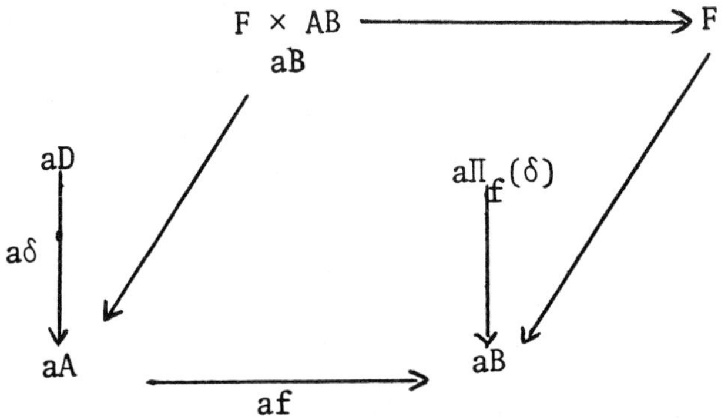

where $f, \delta \in \mathcal{A}$ and $F \in |\neg\neg\mathcal{A}|$.

Since $F = \lim_{\to \mathcal{A}/F} aC$, we can derive the general case from $F = aC$. The trouble here is that we have morphisms $aC \to aB \in \neg\neg\mathcal{A}$ and not just $C \to B \in \mathcal{A}$. The following result, however, takes care of this difficulty (see II.2.13 for notations):

LEMMA 3.2: *If* $A \in |\mathcal{A}|$, *then* $aA = LA$, *where*

$$LA(-) = \lim_{\to R \in J(-)} {}^{opp}\hat{\mathcal{A}}(R, A).$$

Since a morphism $aC \to aB \in \neg\neg\mathcal{A}$ gives rise to some $C \to aB \in \hat{\mathcal{A}}$ (by composition with $C \to aC$), this lemma gives us some $R' \to B \in \hat{\mathcal{A}}$, with $R' \in J(C)$.

Assume now that we are given $aC \to a\Pi_f(\delta) \in \neg\neg\mathcal{A}/aB$. In the same way as before, we obtain some $R'' \to \Pi_f(\delta) \in \hat{\mathcal{A}}/B$ with $R'' \in J(C)$. In other words, we have obtained a diagram

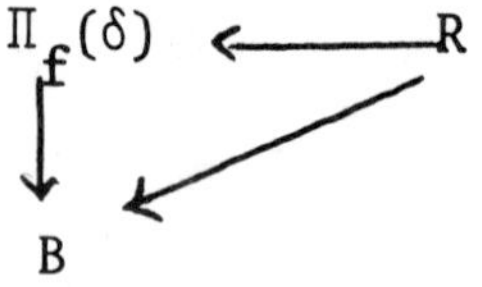

where $S = R' \cap R'' \in J(C)$. We obtain thus two arrows from S to B which are co-equalized by $B \to aB$. Their equalizer R belongs then to $J(C)$.

By III.1.5, we obtain a unique morphism $R \times_B A \to D \in \hat{\mathcal{A}}/A$ which gives us, via the functor a, some $aR \times aA \to aD \in \neg\neg\mathcal{A}/aA$. Since $aR = aC$, as can be checked, we finally obtain a morphism $aC \times_{aB} aA \to aD \in \neg\neg\mathcal{A}/aA$.

The proof is analogous in the case that we start with a morphism $aC \times_{aB} aA \to \in \neg\neg\mathcal{A}/aA$.

Remark 3.3:

(i) More generally, we have proved that for any small site $\mathcal{A}$, the functor $\mathcal{A} \xrightarrow{a} \tilde{\mathcal{A}}$ preserves Π_f whenever it exists, provided that $aA = LA$, for every $A \in |\mathcal{A}|$.

(ii) At the end of this chapter we shall give an example of a small Boolean category for which $aA \not\simeq A$, in general.

PROPOSITION 3.4: $\neg\neg \dashv U$. *More precisely, for every sup-preserving Boolean functor* $\mathcal{A} \xrightarrow{u} \mathcal{B}$ *from a small Boolean category into a Boolean topos* $\mathcal{B}$, *there is an essentially unique morphism of topos* $\neg\neg\mathcal{A} \underset{V}{\overset{U}{\rightleftarrows}} \mathcal{B}$ *such that*

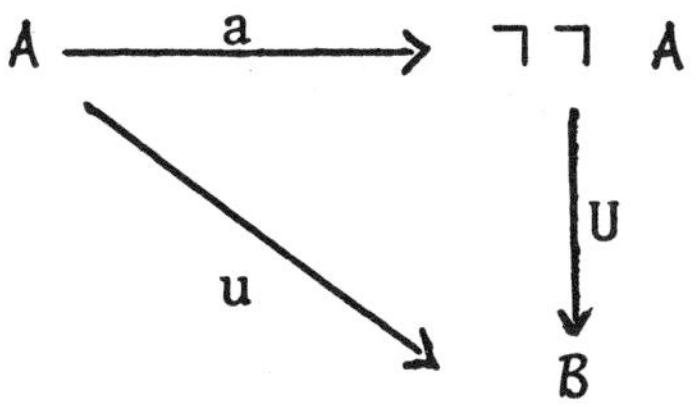

is commutative.

Proof: Exactly as that of the corresponding proposition for the finite cover topology (III.1.7).

Remark 3.5: Let us say that a topos satisfies AC if every epimorphism has a section. It is not true, in general, that $\neg\neg\mathcal{A}$ satisfies AC, as the following shows:

PROPOSITION: *Let* $\mathcal{A}$ *be a small Boolean category. Then the following are equivalent:*

(1) $\neg\neg\mathcal{A}$ *satisfies* AC,
(2) $P(1)$ *is a set of generators for* $\mathcal{A}$.

Condition (2) is a consequence of the projectivity of 1 in $\mathcal{A}$; however it is weaker as shown by the following construction of a Boolean category $\mathcal{A}_0$ such that 1 is not projective though $P(1)$ is a set of generators of $\mathcal{A}_0$.

Let L be a first-order language whose only non-logical symbols are a unary predicate symbol P and a countable set of constants $a_0, a_1, a_2, \ldots$.

Let T be the theory in the language L whose axioms express that the a_n's are different for different n's and that both P and $\sim P$ are infinite.

Notice that T admits elimination of quantifiers, i.e., for every formula $\Phi(x_1, \ldots, x_n)$ there is a quantifier-free formula $\Phi_0(x_1, \ldots, x_n)$ such that $T \vdash \forall x_1, \ldots, \forall x_n(\Phi \leftrightarrow \Phi_0)$.

We let $\mathcal{A}_0$ be the Boolean category obtained from T, as sketched in the introduction.

We first show that $P(1)$ is a set of generators. In this Boolean context, this means that for every $A \neq 0_A$ in $\mathcal{A}_0$ there is some $s: U \to A$ for some $U \rightarrowtail 1$ such that $U \neq 0_1$. Let $A \neq 0_A$ in $\mathcal{A}_0$. Then A can be represented by a formula $\Phi(P, a_1, \ldots, a_n, x_1, \ldots, x_m)$ such that $T \cup \{\exists x_1 \ldots \exists x_m \Phi\}$ is consistent. Since T admits elimination of quantifiers this last condition can be reformulated thus: there are some constants $a_{k_1}, \ldots, a_{k_m}$ such that

$$T \cup \{\Phi(a_{k_1}/x_1, \ldots, a_{k_m}/x_m)\}$$

is consistent. Let $U = [\Phi(a_{k_1}/x_1, \ldots, a_{k_m}/x_m)]$ be the equivalent class of this sentence. Then $U \rightarrowtail 1 \in \mathcal{A}_0$ and $U \neq 0_1$. Moreover, the formula

$$\Phi(a_{k_1}/x_1, \ldots, a_{k_m}/x_m) \wedge x_1 = a_{k_1} \wedge \cdots \wedge x_m = a_{k_m}$$

defines a morphism $s: U \to A$.

From the rather trivial nature of T, it is clear that for no $\theta(x_1)$, $T \vdash \exists ! x_1(\theta(x_1) \wedge Px_1)$, although we have $T \vdash \exists x_1 Px_1$. Translated into statements about $\mathcal{A}_0$ and letting $P = [Px_1]$ be the equivalent class of Px_1, this means that the regular epimorphism $P \to 1 \in \mathcal{A}_0$ has no section. We have thus shown that 1 is not projective.

At the same time, this example shows that P (considered as a representable functor) is not a sheaf for the double negation topology on $\mathcal{A}_0$. Indeed, let us define $\sigma_1 = Pa_1$, $\sigma_{n+1} = Pa_{n+1} \wedge \neg(\sigma_1 \vee \cdots \vee \sigma_n)$ and let $U_n = [\sigma_n] \in |\mathcal{A}_0|$. Then the family $\{U_n \rightarrowtail 1\}_{n \in N}$ is a covering of 1 for the double negation topology.

Letting $U_n \xrightarrow{\xi_n} P \in \mathcal{A}_0$ be the morphism defined by the functional relation $\sigma_n \wedge x_1 = a_n$, the family $(\xi_n)_{n\in N}$ belongs to

$$\mathrm{Eq}\,(\prod_{n\in N} P(U_n) \rightrightarrows \prod_{n,m\in N} P(U_n \cap U_m)).$$

However, as already observed, there is no $\xi: 1 \to P$ and hence P is not a sheaf.

SECTION 4

COMPLETENESS

We now restrict our attention to Boolean categories. The main results of this section are: a description of models as limit ultrapowers (§1), the completeness theorem for Boolean categories (§2) and a reformulation of Robinson-Zakon's notion of enlargement in this context (§3).

1. MODELS AS LIMIT ULTRAPOWERS

Let $\mathfrak{A}$ be a Boolean category and let Φ be an ultrafilter of $P(A)$, for $A \in |\mathfrak{A}|$.

DEFINITION 1.1: We define $h^\Phi: \mathfrak{A} \to \mathcal{S}$ as follows:

$$h^\Phi = \lim\nolimits_{\to B\in\Phi} h^B.$$

A functor $F: \mathfrak{A} \to \mathcal{S}$ is *ultra-representable* if for some $A \in |\mathfrak{A}|$ and some ultrafilter Φ on $P(A)$, F is isomorphic to h^Φ.

A more explicit description of h^Φ is obtained by returning to $\lim_\to$ in $\mathcal{S}$. In fact,

$$h^\Phi(C) = (\coprod_{B\in\Phi} \mathfrak{A}(B, C))/\sim,$$

where the equivalence relation $\sim$ is defined by: $f \sim g$ iff there is

$B' \in \Phi$ such that the diagram

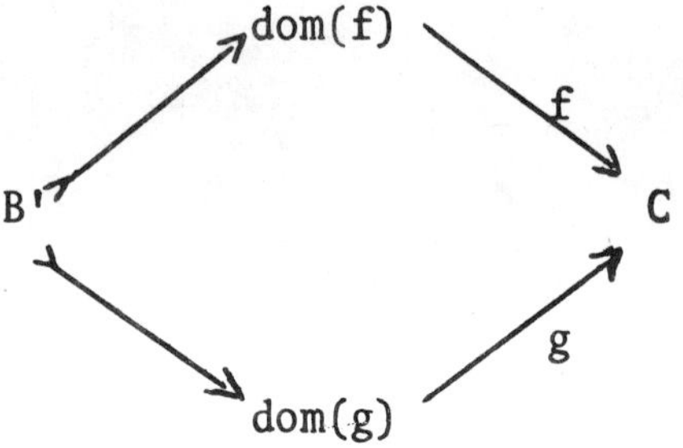

is commutative.

The following result is essentially due to Łos:

THEOREM 1.2: *Every ultrarepresentable functor preserves finite* $\lim_{\leftarrow}$, $\vee$ *and* $\neg$. *Furthermore, if every sub-object of A is projective for regular epimorphisms and Φ is an ultrafilter on $P(A)$, then h^{Φ} is a Boolean model.*

Proof: Since filtered $\lim_{\rightarrow}$ commute with finite $\lim_{\leftarrow}$ in $\mathcal{S}$, we are reduced to show that the supremum of two sub-objects of an object is preserved.

Assume that $F \underset{\sim}{\rightarrow} h^{\Phi}$ for some ultra-filter Φ on $P(A)$. Let $B, C \in P(D)$ and $A' \xrightarrow{\phi} B \vee C \in \mathcal{A}$ for some $A' \in \Phi$. Since $\vee$ are stable, $A' = \phi^{-1}(B) \vee \phi^{-1}(C)$. By definition of an ultra-filter, either $\phi^{-1}(B) \in \Phi$ or $\phi^{-1}(C) \in \Phi$. Assume the first case. From the pull-back diagram

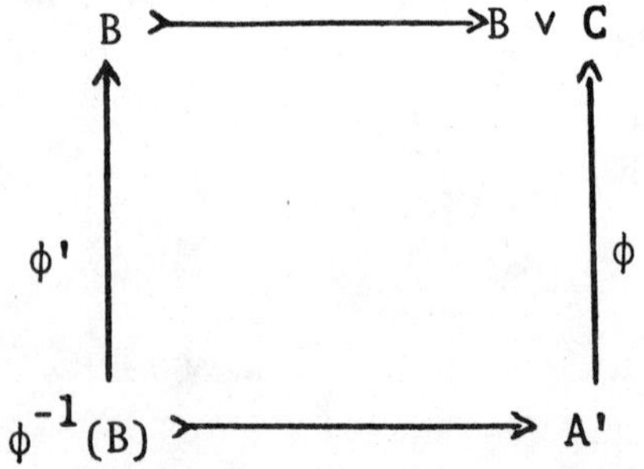

we conclude that $\phi' \sim \phi$, i.e., $\phi/\sim \in h^{\Phi}(B)$. This implies that $h^{\Phi}(B \vee C) = h^{\Phi}(B) \cup h^{\Phi}(C)$.

To prove the second part, we have to check that h^{Φ} preserves regular epimorphisms. Since these are co-equalizers, i.e., right limits, we are reduced to show that h^{B} preserves regular epimorphisms, for $B \in \Phi$. Let $C \overset{g}{\twoheadrightarrow} D$ be regular and $B \overset{\phi}{\longrightarrow} D \in \mathfrak{A}$. Since regular epimorphisms are stable, the left vertical arrow in the pull-back diagram

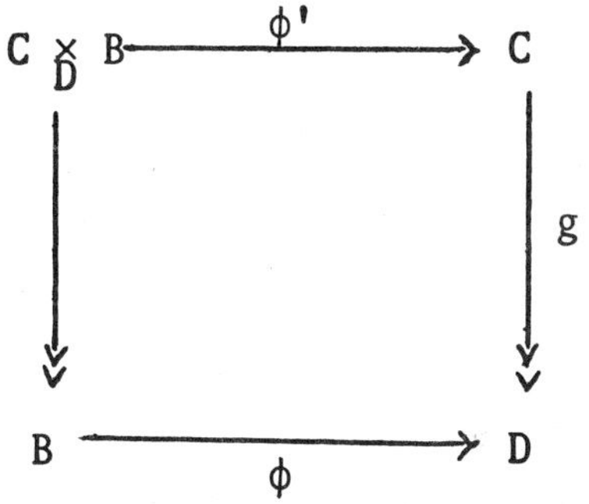

is again regular epic. By assumption, it has a section $B \overset{s}{\longrightarrow} C \underset{D}{\times} B$.

Therefore the diagram

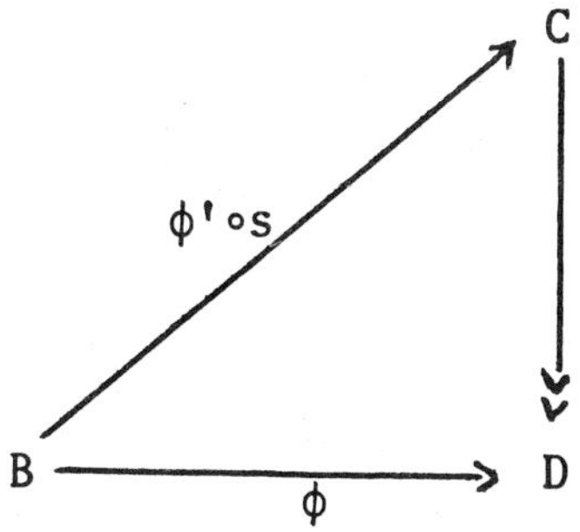

is commutative and this shows that $h^{B}(C) \to h^{B}(D)$ is epic.

COROLLARY 1.3: *Every filtered* $\lim_{\rightarrow}$ *of ultra-representable functors preserves finite* $\lim_{\leftarrow}$, $\vee$ *and* $\neg$. *Furthermore, if every regular epimorphism has a section, the filtered* $\lim_{\rightarrow}$ *of ultra-representable functors are Boolean models.*

The converse of this corollary is true. It is essentially a reformulation of results of Engeler, Keisler, Kochen on limit ultrapowers and follows the lines of Joyal [J2].

THEOREM 1.4: *Let* $\mathcal{A}$ *be a small Boolean category. Then every functor into* $\mathcal{S}$ *which preserves finite* $\lim_{\leftarrow}$, $\vee$ *(and* $\neg$*) is a filtered* $\lim_{\rightarrow}$ *of ultra-representable functors.*

Proof: Let $\mathcal{A}$ be a small Boolean category and let $M: \mathcal{A} \rightarrow \mathcal{S}$ be a functor which preserves finite $\lim_{\leftarrow}$ and $\vee$.

We shall consider the "comma category" $M/\mathcal{A}$ whose objects are formal arrows $M \xrightarrow{\xi} A$ and the morphism triangles

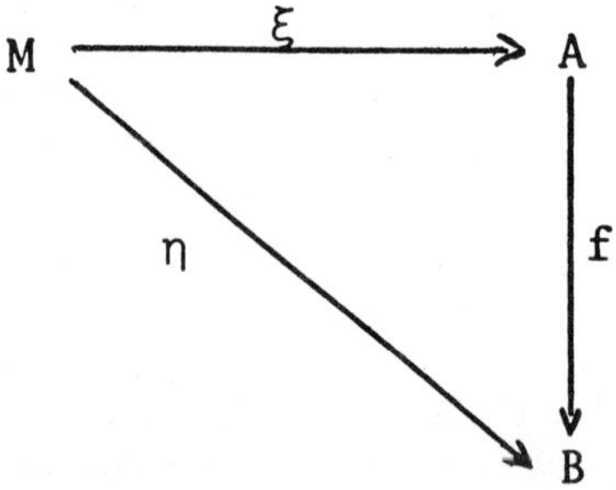

such that $\eta = M(f)(\xi)$.

Since $\mathcal{A}$ has finite limits and M commutes with these, $M/\mathcal{A}$ has finite limits again and its dual $I = (M/\mathcal{A})^{op}$ is a small filtered category.

We now define a functor from I into the category of ultra-representable functors of $\mathcal{A}$ as follows: if $M \xrightarrow{\xi} A \in |I|$, let $\Phi(\xi) = \{B \rightarrowtail A : \xi \in M(B)\}$. It is easily checked (using the fact that M is a model) that $\Phi(\xi)$ is a proper ultra-filter of the Boolean algebra $P(A)$. We associate with $M \xrightarrow{\xi} A$ the ultra-representable functor $h^{\Phi(\xi)}$. This association is functorial. Furthermore, for each $M \xrightarrow{\xi} A$ we can define a natural transformation $h^{\Phi(\xi)} \xrightarrow{\varphi_\xi} M$ by $\phi_\xi(C)(\phi) = M(\phi)(\xi)$, where $A' \xrightarrow{\phi} C$ and $A' \in \Phi(\xi)$. Notice that this last condition is equivalent to $\xi \in M(A')$, hence $M(\phi)(\xi) \in M(C)$ and $\phi_\xi(C)$ is well-defined.

Since M preserves equalizers, ϕ_ξ is easily seen to be a monomorphism. To finish the proof, we check that $M \cong \lim_{\to \xi \in I} h^{\Phi(\xi)}$. In fact, let $\xi \in M(A)$. Consider the map $h^{\Phi(\xi)}(A) \xrightarrow{\phi_\xi(A)} M(A)$. The value of $\phi_\xi(A)$ at 1_A is given by $\phi_\xi(A)(1_A) = M(1_A)(\xi) = \xi$.

2. COMPLETENESS THEOREM FOR BOOLEAN CATEGORIES

We now sketch Joyal's proof of the completeness theorem.

THEOREM 2.1: *Let $\mathcal{A}$ be a small Boolean category. If A is distinct from the smallest object of $P(A)$, then there is a Boolean model $M: \mathcal{A} \to \mathcal{S}$ such that $M(A) \neq \varnothing$.*

Proof: Assume first that $\mathcal{A}$ has finite sums. In this case, we can show (either directly or applying II.2.6 to $\tilde{\mathcal{A}}$, the category of sheaves for the finite cover topology and "reflecting" back these properties to $\mathcal{A}$) that the sums are universal, the canonical injections $A_i \to \coprod_{i \in I} A_i$ are monomorphisms and $A_i \underset{S}{\times} A_j$ is initial for $i \neq j$, $S = \coprod_{i \in I} A_i$. We shall use the following two simple results.

LEMMA 2.2: *If $\mathcal{A}$ is a Boolean category and $A \in |\mathcal{A}|$, then $\mathcal{A}/A$ is again a Boolean category. Furthermore, if $\mathcal{A}$ has finite sums, so does $\mathcal{A}/A$.*

LEMMA 2.3: *If $\mathcal{A}$ is a Boolean category and $A \xrightarrow{f} B \in \mathcal{A}$, then the pulling-back functor $\mathcal{A}/B \xrightarrow{*f} \mathcal{A}/A$ is Boolean. Furthermore, $*f$ is faithful, whenever f is a regular epimorphism.*

Let $\{A_i \overset{\Pi_i}{\twoheadrightarrow} 1\} : i \in I$ be a list of all regular epimorphisms into the final object 1. If $F \subset G \subset I$ are finite subsets of I, we let

$$\prod_{I \in G} B_i \xrightarrow{\Pi_F^G} \prod_{I \in F} B_i$$

be the canonical projection. Since Π_F^G is regular epic (by stability of regular epimorphisms),

$$\mathcal{A}/\prod_{i \in F} B_i \xrightarrow{{}^*\Pi_F^G} \mathcal{A}/\prod_{i \in G} B_i$$

is a faithful Boolean functor (previous lemmas). We now let $\mathcal{A}_0 = \lim_{\to F \subset I} \mathcal{A}/\prod_{i \in F} B_i$. It is easily checked that $\mathcal{A}_0$ is again a Boolean category. Since $\mathcal{A} \overset{\sim}{\to} \mathcal{A}/1$, we have a faithful Boolean functor $\mathcal{A} \to \mathcal{A}_0$. Repeating this construction ω times, we obtain a sequence

$$\mathcal{A} \to \mathcal{A}_0 \to \mathcal{A}_1 \to \cdots .$$

whose $\lim_{\to}$ is a Boolean category $\mathcal{A}_\infty$.

We claim that every sub-object of 1 in $\mathcal{A}_\infty$ is projective for regular epimorphisms. In fact, from the construction of $\mathcal{A}_\infty$ it is easy to see that 1 is projective for regular epis. Let $A \overset{f}{\twoheadrightarrow} B$ be a regular epimorphism and let $B \rightarrowtail 1$. If $\neg B \rightarrowtail 1$ is the complement of $B \rightarrowtail 1$ in $P(1)$ and $A \coprod \neg B$ is the disjoint sum, then the canonical injections $A \to A \coprod \neg B$, $\neg B \to A \coprod \neg B$ are monomorphisms and we have the following commutative diagram

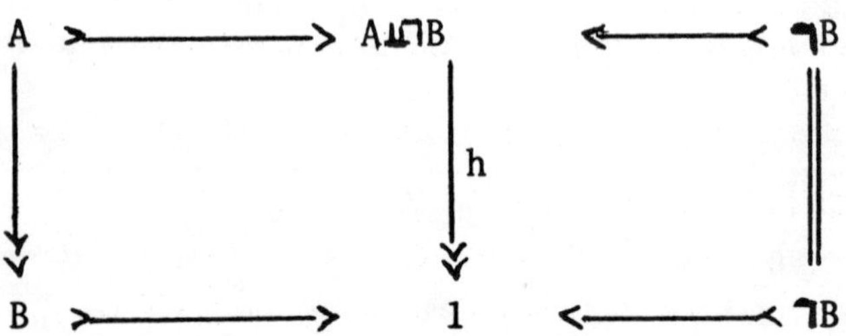

It is easy to see that h is a regular epimorphism and so it has a section S. By the universality of sums,

$$1 = S^{-1}(A \coprod \neg B) = S^{-1}(A) \coprod S^{-1}(\neg B).$$

Since the diagram

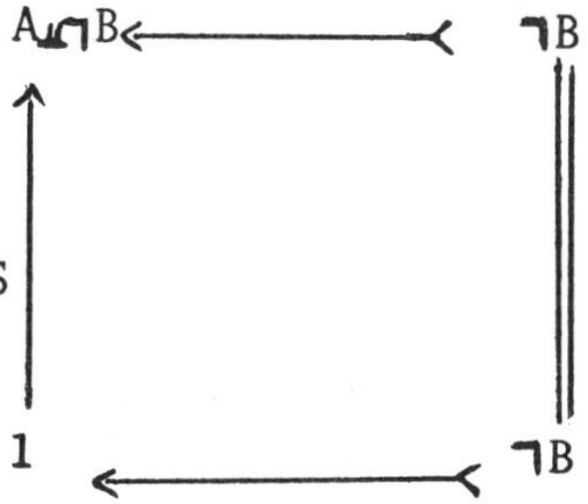

is a pull-back, $S^{-1}(\neg B) = \neg B$. This implies that $B \rightarrowtail S^{-1}(A)$. Since $\exists_S$ is a left adjoint to S^{-1}, $\exists_S(B) \rightarrowtail A$ and this shows that $B \rightarrowtail 1 \xrightarrow{s} A \coprod \neg B$ factors through A, i.e., f has a section.

Let us come back to the proof of the completeness theorem. Assume that $A \neq 0_A$, the smallest object of $P(A)$. Then $A \neq 0_A$ in $\mathcal{A}_\infty$, since $\mathcal{A} \to \mathcal{A}_\infty$ is faithful. Let $A \twoheadrightarrow A' \rightarrowtail 1$ be the unique factorization of $A \to 1$ in a regular epimorphism followed by a monomorphism in $\mathcal{A}_\infty$. Since $0 \rightarrowtail 1$ is strict, $A' \neq 0$. Let Φ be any proper ultrafilter on $P(1)$ containing A'. Then h^Φ is a model of $\mathcal{A}_\infty$ (IV.1.2.) and $h^\Phi(A) \neq \varnothing$.

If $\mathcal{A}$ does not have finite sums, we first consider the category $\neg\neg\mathcal{A}$ of sheaves for the double negation topology. The functor $\mathcal{A} \xrightarrow{a} \neg\neg\mathcal{A}$ is Boolean (III §3), furthermore it is easy to see that a is faithful.

Let $\mathcal{A}'$ be any small Boolean full sub-category of $\neg\neg\mathcal{A}$ such that

(i) $|\mathcal{A}'| \supset |a(\mathcal{A})|$ and
(ii) $\mathcal{A}'$ has finite sums.

We thus obtain a faithful Boolean functor $\mathcal{A} \rightarrowtail \mathcal{A}'$.

Hence $A \neq 0_A$ in $\mathcal{A}'$ and, by the previous argument, we can find a model $M: \mathcal{A}' \to \mathcal{S}$ such that $MA \neq \varnothing$. The functor

$$\mathcal{A} \rightarrowtail \mathcal{A}' \xrightarrow{M} \mathcal{S}$$

is again a model with the same property.

3. NON-STANDARD ANALYSIS

Following Robinson-Zakon [RZ], we shall define the superstructure $\mathcal{A}$ generated by a set X by recursion:

$$\mathcal{A}_0 = X$$

$$\mathcal{A}_{n+1} = P\left(\bigcup_{i=0}^{n} \mathcal{A}_i\right)$$

$$\mathcal{A} = \bigcup_{n=1}^{\infty} \mathcal{A}_n.$$

In an obvious way, $\mathcal{A}$ can be considered as a full sub-category of the category of sets $\mathcal{S}$. Then $\mathcal{A}$ is a Boolean category having the property that every epimorphism has a section (axiom of choice).

Let $M: \mathcal{A} \to \mathcal{S}$ be any model (i.e., Boolean functor) of $\mathcal{A}$. If $A \in |\mathcal{A}|$ and ψ is any filter-base of subsets of A, we let $\mathrm{Ker}_M(\psi) = \bigcap_{B \in \psi} M(B)$.

We say that M is an *enlargement* of $\mathcal{A}$ if for every $A \in |\mathcal{A}|$ and every filter base ψ on A, $\mathrm{Ker}_M(\psi) \neq \varnothing$.

THEOREM 3.1: *Every super-structure has enlargements.*

Proof: We shall consider the set $\mathfrak{F}$ of couples (A, ψ) such that $A \in |\mathcal{A}|$ and ψ is a filter-base on A and let

$$I = \prod_{(A,\psi) \in F} A.$$

We let $\pi_{(A,\psi)}: I \to A$ be the canonical projections. Since

$$\{\pi_{(A,\psi)}{}^{-1}(B) : B \in \psi, (A, \psi) \in \mathfrak{F}\}$$

is a filter-base on I, we can extend it to some proper ultrafilter Φ. We claim that the restriction to $\mathfrak{A}$ of the ultra-representable functor h^{Φ} is an enlargement of $\mathfrak{A}$. In fact, let $(A, \Psi) \in \mathfrak{F}$ and let $B \in \Psi$. Then $\pi_{(A,\psi)}{}^{-1}(B) \in \Phi$ and this shows that the restriction of $\pi_{(A,\psi)}$ to some member of Φ factors through B, i.e., the canonical image of $\pi_{(A,\psi)}$ in $h^{\Phi}(A)$ belongs to $h^{\Phi}(B)$. The rest of the proof follows from IV 1.2.

As an application, let $\mathfrak{A}$ be the super-structure generated by the field of reals $\mathbb{R}$ and let $M: \mathfrak{A} \to \mathfrak{S}$ be a Boolean model.

We claim that $M(\mathbb{R})$ is again a field. The associativity of $+$, for instance, is equivalent to the commutativity of the following diagram

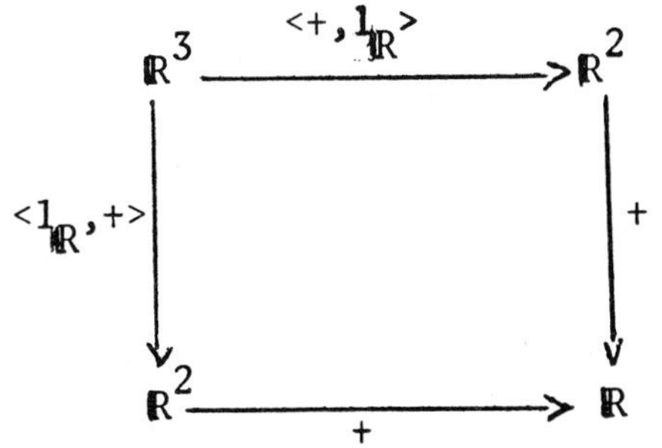

Since M preserves products,

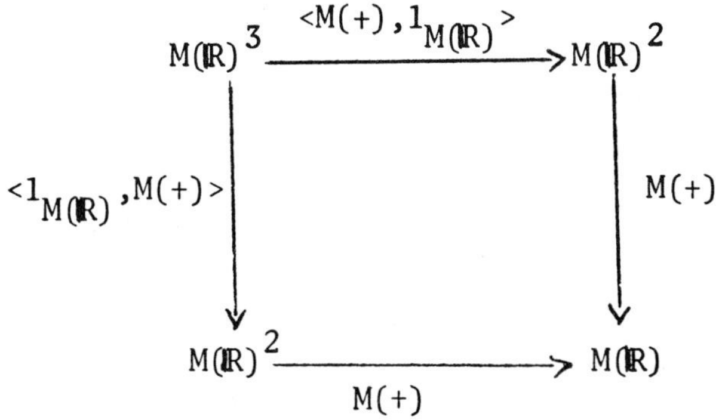

is again commutative and this means that $M(+)$ is associative.

To check that every non-zero element in $M(\mathbb{R})$ has an inverse, we consider the commutative diagram

$$\begin{array}{ccccc} R & \xleftarrow{\pi_1} & R^2 & \xrightarrow{\ \cdot\ } & R \\ \uparrow & & \uparrow & & \uparrow \\ R-\{0\} & \xleftarrow[p]{} & \cdot^{-1}\{1\} & \longrightarrow & \{1\} \end{array}$$

The fact that $\mathbb{R}$ itself has this property is equivalent to saying that $Im(p) = \mathbb{R} - \{0\}$. Now, M preserves fibered products and the right square is a pull-back. Furthermore M preserves Boolean operations and images and so $Im(M(p)) = M(\mathbb{R}) - \{0\}$, i.e., $M(\mathbb{R})$ is a field.

Of course, there is a more conventional (and perhaps more efficient) way of stating the preservation properties of a model. For this, we define the elementary properties of $\mathfrak{A}$ in terms of a first-order language with bounded quantifiers (see [RZ] for details). However, this algebraic language-free approach may be more palatable to some.

We conclude with another characterization of enlargements.

Let $\mathcal{K}$ be the category of compact (not necessarily Hausdorff) topological spaces and open maps (as morphisms) and let U: $\mathcal{K} \to \mathcal{S}$ be the forgetful functor into the category of sets.

Proposition 3.2: *Let $\mathfrak{A}$ be a super-structure and let $M: \mathfrak{A} \to \mathcal{S}$ be a Boolean model. Then the following are equivalent:*

(1) *M is an enlargement,*
(2) *M is faithful and factors through U.*

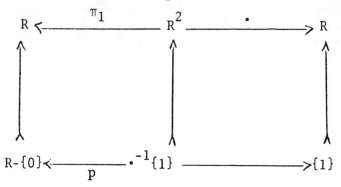

Proof: (1) $\Rightarrow$ (2): For every $A \in |\mathfrak{A}|$, we introduce a topological structure on $M(A)$ by letting $\{M(B) : B \subset A\}$ be a basis for the topology. Since $M(A \sim B) = M(A) \sim M(B)$, $M(A)$ has a basis of clopen sets. It is enough to check compactness for the basis. Let $(M(B_i))_{i\in I}$ be such that $\bigcap_{i\in F} M(B_i) \neq \varnothing$ for every finite $F \subset I$. Then $\Psi = \{B_i: i \in I\}$ is a filter-base on A and thus

$$\bigcap_{i\in I} M(B_i) = \mathrm{Ker}_M(\Psi) \neq \varnothing.$$

(2) $\Rightarrow$ (1): Assume that we have a commutative diagram

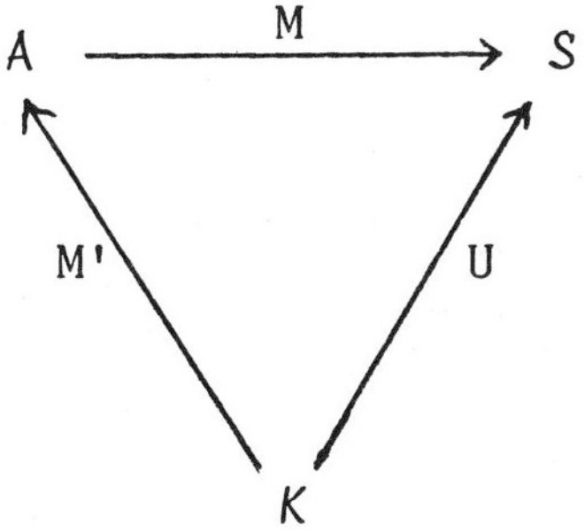

For every (A, Ψ), we claim that $\mathrm{Ker}_M(\Psi) \neq \varnothing$. In fact, for all $B \in \Psi$, $M(B) \subset M(A)$ is closed, indeed clopen. Furthermore

$$M(B_1) \cap \cdots \cap M(B_n) = M(B_1 \cap \cdots \cap B_n) \neq \varnothing$$

for every $B_1, \ldots, B_n \in \Psi$, since M is faithful. By compactness $\mathrm{Ker}_M(\Psi) \neq \varnothing$.

REFERENCES

B Barr, M., et al., *Exact Categories and Categories of Sheaves.* Lecture Notes in Mathematics 236, Springer-Verlag, 1971.

J1 Joyal, A., "Polyadic spaces and elementary theories," Notices AMS vol. 18 (1971), no. 3, 563.

J2 ———, "Functors which preserve elementary operations," Notices AMS vol. 18 (1971), no. 6, 967.

KW Kock, A., and G. C. Wraith, "Elementary Toposes," Lecture Notes Series No. 30, Aarhus University, 1971.

L1 Lawvere, F. W., "Functorial semantics of elementary theories," JSL **31** (1966), 294.

L2 ———, "Theories as categories and the completeness theorem," JSL **32** (1967), 562.

L3 ———, "Quantifiers and Sheaves," Actes, Congrès Internat. Math., (1970) (Nice) 1–6.

P Pareigis, B., *Categories and Functors*. New York: Academic Press, 1970.

RS Rasiowa, H., and R. Sikorski, *The Mathematics of Metamathematics*. Warsaw: PAN, 1963.

RZ Robinson, A., and E. Zakon, "A set-theoretical characterization of enlargements," in *Applications of Model Theory to Algebra, Analysis, and Probability*. New York: Holt, Rinehart and Winston, 1969.

S Solovay, R. M., "New proof of a theorem of Gaifman and Hales," *Bull. Amer. Math. Soc.*, **72** (1966), 282–284.

V Verdier, J. L., "Topologies et Faisceaux," Fascicule 1, Sém. de Géom. Algébrique 63–64, Institut des Hautes Etudes Scientifiques.

V1 Volger, H., "Logical Categories," Dalhousie University, 1971.

V2 ———, "Completeness theorem for logical categories," to appear.

INDEX